A Tinkerer's Guide to CNC Basics

Master the fundamentals of CNC machining, G-Code,
2D Laser machining and fabrication techniques

Samer Najia

BIRMINGHAM—MUMBAI

A Tinkerer's Guide to CNC Basics

Copyright © 2024 Packt Publishing

Group Product Manager: Rohit Rajkumar

Publishing Product Manager: Vaideeshwari Muralikrishnan

Book Project Manager: Aishwarya Mohan

Senior Editor: Rakhi Patel

Technical Editor: Simran Udasi

Copy Editor: Safis Editing

Proofreader: Safis Editing

Indexer: Pratik Shirodkar

Production Designer: Alishon Mendonca

DevRel Marketing Coordinators: Namita Velgekar and Nivedita Pandey

First published: February 2024

Production reference: 1040124

Published by Packt Publishing Ltd.

Grosvenor House

11 St Paul's Square

Birmingham

B3 1RB, UK

ISBN 978-1-80324-749-6

www.packtpub.com

To my wife, Sanja, for lovingly and patiently putting up with my projects and incessant desire to build things. To my children, Hanna and Jordan, for inspiring my creativity.

– Samer Najia

Contributors

About the author

Samer Najia has always enjoyed building things and often has multiple projects in the air. While by day he is in the IT field, at all other times he is often putting bits of things together in the garage or on his desk. When the sky beckons, Samer likes to fly and work on airplanes.

About the reviewer

Atif Tajul, born in Kuala Lumpur, Malaysia, is a mechanical engineer who graduated from the University of Southampton, United Kingdom. He indulges himself in anything hands-on and enjoys tinkering. He is a very practical person when solving problems and providing solutions. He has experience of working on cars as an apprentice mechanic, and the skills and knowledge he gains from that are used for the benefit of his family. Atif 3D prints personal designs with his own machine in the pursuit of becoming a master at prototyping and fabrication. He keeps himself in shape by playing football as he is a massive fan of the sport.

Table of Contents

7

Enclosures   77

8

Project: Building a CNC Laser Cutter and a Plotter   85

9

Project: Building Your Own 4th Axis   97

10

Project: Adding a Laser to the 3018   111

11

Building a More Capable CNC Machine   121

12

Future Projects and Going Bigger and Better   131

Preface

A Tinkerer's Guide to CNC Basics will suit anyone who enjoys shop work and tinkering through the process of automating the fabrication of parts of various materials, including cutting and engraving with milling machines and lasers. If you have a desire to make things out of wood, metal, plastic, foam, fiberglass, or other materials, and maybe have to make several parts repeatedly, this book is for you. If you need to prototype your designs and want to be able to do so fast so you don't have to hand fabricate everything and learn how to leverage **Computer-Aided Design** (**CAD**), you will benefit from this book.

The book starts with an overview of what CNC is and progresses toward acquiring, building, and customizing a commonly used CNC machine before delving into various projects. These projects include upgrades to this machine, building larger and more complex machines, and fabricating parts for specific applications. You will learn how to operate and service a desktop CNC machine, use CAD to design or modify parts that your machine can then fabricate, and finally learn how to scale up your efforts with bigger and more complex systems.

By the time you are finished with this book, you will know how to fabricate using a basic CNC machine, cut with a laser, use a fourth axis to cut parts as they rotate on your work table, and operate multiple software applications to achieve your desired outcomes. You will also become familiar with a number of techniques to transfer drawings from paper to electronic formats suitable for fabrication by your CNC mill.

Who this book is for

Anyone who enjoys working in their home shop or garage or likes to tinker and build things from scratch out of many materials will enjoy this book. Tinkerers will grow their skills and add automation to their repertoire of tools to fabricate just about anything.

What this book covers

Chapter 1, The What and Why of CNC: This chapter introduces CNC, discusses the mechanics of how it works, and provides some initial considerations for safety.

Chapter 2, Setting Up and Configuring the 3018 CNC Machine: This chapter's focus is on the build-or-buy decision and getting your first CNC machine up and running.

Chapter 3, Understanding Material Properties before Making the First Cut: Here, we take a look at what is needed for various materials as far as the CNC machine is concerned, such as how to select an endmill for a particular application.

Chapter 4, Making the First Cut: With this chapter, we'll put the 3018 to work and start cutting materials.

Chapter 5, Full CNC Workflow with Different Materials: Building on the previous chapter, we begin looking at how to go from design to finished product, including the transfer of paper drawings to an electronic format suitable for subsequent processing with our 3018 machines.

Chapter 6, Upgrading Your CNC Machine: We'll add components to the machine we have to be more precise, start with a discussion on a fourth axis, and add the ability for our machine to become a plotter and a drag knife.

Chapter 7, Enclosures: CNC machines produce debris, and if using a laser, there could be fumes that might need ventilation. This chapter discusses some simple enclosures that can be built to keep your work area neat.

Chapter 8, Project: Building a CNC Laser Cutter and Plotter: Taking everything we have learned so far, we'll build limited-purpose CNC machines: one to generate and scale drawings and another to cut using a laser.

Chapter 9, Project: Building Your Own 4th Axis: In this chapter, we build on *Chapter 6* and build a fourth axis add-on using our 3018 and some off-the-shelf parts.

Chapter 10, Adding a Laser to the 3018: We'll add a laser toolhead to our original desktop CNC mill to make it a 2-in-1 machine.

Chapter 11, Building a More Capable CNC Machine: Once we outgrow the 3018 machine, we will want something bigger and stronger with a larger workspace. This chapter steps through the process of scaling up.

Chapter 12, Future Projects and Going Bigger and Better: We'll look at even bigger machines for our shop including stepping out of the hobbyist arena and seeing what industrial CNC machines can do. We also have a quick look at five-axis CNC machines.

To get the most out of this book

You will need some basic tools to assemble your machines, including drills, drill bits, screwdrivers, hex keys, rulers, and tape measures, some scrap material to use when test cutting with your CNC machine, and a suitable work area. Safety gear is also highly recommended, including eye protection, and when using the laser, special eye protection is mandatory.

Get in touch

Feedback from our readers is always welcome.

General feedback: If you have questions about any aspect of this book, mention the book title in the subject of your message and email us at customercare@packtpub.com.

Errata: Although we have taken every care to ensure the accuracy of our content, mistakes do happen. If you have found a mistake in this book, we would be grateful if you would report this to us. Please visit www.packtpub.com/support/errata, select your book, click on the Errata Submission Form link, and enter the details.

Piracy: If you come across any illegal copies of our works in any form on the Internet, we would be grateful if you would provide us with the location address or website name. Please contact us at copyright@packt.com with a link to the material.

If you are interested in becoming an author: If there is a topic that you have expertise in and you are interested in either writing or contributing to a book, please visit authors.packtpub.com.

Reviews

Please leave a review. Once you have read and used this book, why not leave a review on the site that you purchased it from? Potential readers can then see and use your unbiased opinion to make purchase decisions, we at Packt can understand what you think about our products, and our authors can see your feedback on their book. Thank you!

For more information about Packt, please visit packtpub.com.

Share Your Thoughts

Once you've read *A Tinker's Guide to CNC Basics*, we'd love to hear your thoughts! Scan the QR code below to go straight to the Amazon review page for this book and share your feedback.

https://packt.link/r/1803247495

Your review is important to us and the tech community and will help us make sure we're delivering excellent quality content.

Download a free PDF copy of this book

Thanks for purchasing this book!

Do you like to read on the go but are unable to carry your print books everywhere?

Is your eBook purchase not compatible with the device of your choice?

Don't worry, now with every Packt book you get a DRM-free PDF version of that book at no cost.

Read anywhere, any place, on any device. Search, copy, and paste code from your favorite technical books directly into your application.

The perks don't stop there, you can get exclusive access to discounts, newsletters, and great free content in your inbox daily

Follow these simple steps to get the benefits:

1. Scan the QR code or visit the link below

https://packt.link/free-ebook/9781803247496

2. Submit your proof of purchase
3. That's it! We'll send your free PDF and other benefits to your email directly

1

The What and Why of CNC

Computer numerical control (**CNC**) is a software-based method of moving tools and machinery. This method has traditionally been under the purview of factories and manufacturing facilities and well beyond the reach of a garage tinkerer. CNC machines have arms and carriages that can hold mills, grinders, lasers, and other cutting tools that move in multiple axes to shape an object via preprogrammed movement commands.

Our objective in this chapter is to provide a basic understanding of CNC from a hobbyist's/tinkerer's perspective as we prepare to take a deep dive into obtaining, operating, and customizing our own machine.

In this chapter, we will cover the following topics:

- Branches of CNC machining

- Differing approaches to motion in CNC machines

- How CNC works and when to use it

- A brief overview of G-code, the commands used to operate a CNC machine

- Safety considerations for CNC and lasers

Branches of CNC machining

CNC manufacturing can be traced back to the 1940s, when the first **numerical control** (**NC**) machines started to appear (`https://en.wikipedia.org/wiki/History_of_numerical_control`), and methods to automate handcrafted fabrication can be traced to three centuries ago. I'm sure you've seen videos of these machines perhaps fabricating the engine block for a car or cutting and shaping sheet metal. For all intents and purposes, a CNC machine is a type of robot. It takes stock material (a sheet of wood, a block of aluminum) and turns it into a product ready to be assembled or finished very quickly, very accurately, and, most importantly, repeatedly. CNC machining made fabrication at scale possible.

With the advent of desktop computing, more commonly accessible materials, and smaller, more powerful components, it also became possible to bring CNC machining to the home workshop. Now, anyone with a good guide can put together a robust desk or bench-top CNC machine and get to cutting, engraving, and milling themselves. Such machining is not limited to metal, wood, foam, and other materials; there are multiple branches of CNC machining:

- **PCB milling**: Where holes are drilled in the right places and the circuit board is cut and shaped for whatever they are going into.

- **Rotary engraving**: Where the movement is around an axis instead of along it. Imagine engraving a bottle, glass, or vase.

- **Multi-axis machining**: Where the machine does not just operate in X, Y, and Z, but also rotates the object being worked on to shape all of its sides (this is a typical use case for machining a car's engine block). For an example of this, look at this video: `https://www.youtube.com/watch?v=tO6AGOjBoGo`.

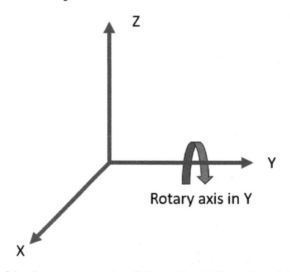

Figure 1.1 – Example of the three main axes in a CNC machine (X, Y, and Z) and the rotary axis around Y

In the preceding diagram, we have the three basic axes common to most CNC machines we will deal with. Horizontally, we have X and Y (lasers, for example, typically cut through, so they do not shape material vertically). The Z axis is for cutting into the material vertically (e.g., when engraving or drilling holes) and when the object being worked on is rotated, we have a rotary axis (here portrayed by turning the object *around* the Y axis). For the other two axes (X and Z), you could also have the part rotated around those axes, but we will not be doing this as the machine complexity increases tremendously. However, we will briefly discuss five-axis machines, which rotate around X and Y and travel along X, Y, and Z, in *Chapter 12, Future Projects and Going Bigger and Better.*

By moving and changing toolheads (a toolhead is the cutting/engraving assembly mounted on the X-carriage of the CNC machine) through automation, the process of crafting complex components is accurate, repeatable, and very scalable. By making design adjustments in software, it also becomes possible to quickly fabricate prototypes before putting the *process* into production.

Differing approaches to motion in CNC machines

Depending on the nature of the machine itself, the motion system can vary based on the rigidity requirements of the toolhead. Very frequently, a compromise is struck between speed and rigidity (or stiffness) so that cheaper or more readily available components may be used. The most common of these tradeoffs is allowing X- and Y-axis motion to be driven by belts while the Z-axis motion operates using a leadscrew. Other designs use leadscrews throughout.

The following figure shows one of my four *3018* machines:

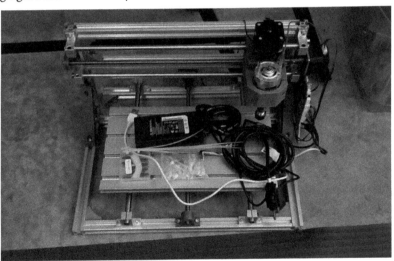

Figure 1.2 – A 3018 machine (see Figure 2.1 for an annotated close-up picture)

In the preceding figure, note the controller on the right. This picture was taken just prior to calibration. This is a machine that can be purchased online as a kit. It is made from some off-the-shelf parts (8020 aluminum extrusions, 8-mm steel rods, stepper motors (which are used for motion), leadscrews, and an engraving or cutting motor, also known as the spindle motor) and some vendor-provided components, such as the plastic parts and the controller board. This design is open source and almost everything can be purchased from various vendors online.

The following figure is of *Bumblebee* (so-called because I 3D-printed parts in yellow plastic and had black anodized aluminum extrusions in my parts bin to make it from):

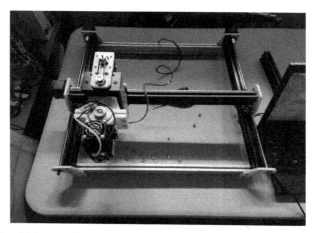

Figure 1.3 – Bumblebee with its dual Y-leadscrews and 2-in-1 toolhead (spindle and laser)

Bumblebee is an example of a motion system that uses Delrin wheels and the extrusions themselves as rails (versus metal rods). This structure is larger than the 3018 machine and so had to be more rigid, which is more difficult when using smooth rods and linear bearings (rods can flex and would need to be thick and heavy).

The CNC machine's toolhead is essentially a motor turning an end-mill or similar bit against a piece of material. That bit is going to experience some resistance and, of course, the more rigid the machine, the higher the likelihood of it being able to handle hard materials. Belts, for example, can stretch under load, which can make the cut less accurate or require multiple passes to handle certain materials. Using belts is typically best when the toolhead is relatively light or will be cutting light materials and is acceptable for lasers since the head only moves above the workpiece.

As an example of a belt-driven system, here is a picture of a laser cutter I built some time ago:

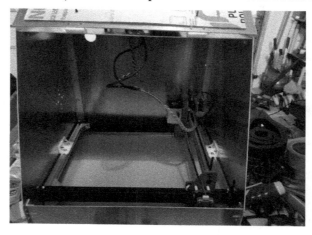

Figure 1.4 – Custom-built 10W laser

The preceding unit is a modification of a kit I had built for a friend, but which proved too underpowered for him. I ended up replacing all the acrylic parts with 3D-printed equivalents, replaced the 500-mW laser with a 10-W blue laser, and installed the entire assembly in a custom aluminum enclosure.

Belt for the X axis of my laser cutter. The belt wraps around a pulley mounted to the motor at back (top red arrow)

Figure 1.5 – Closeup of the X-axis belt on 10-W laser

In the preceding figure, the black line at the end of the red arrows is the belt (recessed into the metal extrusion). This is the gantry that moves in the X axis. There is a similar arrangement for the Y axis.

The reason I put this laser in a metal enclosure is because its previous incarnation jammed on a workpiece that became dislodged and the laser then ignited the workpiece. The flames consumed some of the gantry's plastic parts and wheels. Consequently, I prefer large lasers to be fully enclosed in something that can contain any flames.

As you now have seen, a typical desktop CNC machine essentially moves a toolhead along a cartesian coordinate system (X, Y, Z). However, it is possible to engrave or cut into a workpiece along a rotary axis. In the desktop world, this is commonly achieved by adding a roller onto the work area that rotates the object being worked on in one axis. More often than not, this is in the Y axis. Imagine taking a sheet of paper and rolling it up into a cylinder. Rather than have the gantry move along the length of the paper, we are moving the rolled-up cylinder about its longitudinal centerline as the toolhead cuts into it. Here are some examples of this type of machine that you can make yourself:

Figure 1.6 – An example of a roller axis compatible with desktop laser machines
(Source: Laser Rotary Attachment [https://www.thingiverse.com/thing:5349022];
created by and used with permission from Sean Mullen [zaphod101])

It's not fully apparent in the preceding image, but the horizontal orientation of the cylinder in the pictured example can be adjusted by raising or lowering one side so that the cylinder surface is level. This tactic allows the laser to hit the surface directly overhead and accommodate objects with varying diameters.

The same can be done with a CNC spindle, that is, cutting into an object much like a lathe. Here is another DIY example that we will also explore in *Chapter 6, Upgrading Your CNC Machine*.

Figure 1.7 – A CNC rotary axis (source: Poor Man's 4th Axis cnc [https://www.thingiverse.com/thing:2344975]; created by and used with permission from Daniel Gross [ZenziWerken])

Have a look at the accompanying video located at `https://youtu.be/WfRF8FE9Qgc` to see how the preceding axis is being used to cut treads into a wheel.

It's also important to note that CNC machines typically have very limited motion in the Z (vertical) axis. For most desktop machines, the workpiece is not very tall, and in any case the end-mill or bit would snap from the strain applied to it if it were too long.

How CNC works and when to use it

The CNC machine is operated by an onboard controller that runs three or more stepper motors and the toolhead. Stepper motors have a lot more torque and can be controlled with greater precision than ordinary electric motors. Motor control is measured in fractions of revolutions, which allows for excellent precision in movement. Using precision leadscrews (or pulley/belt systems), the motors move a gantry along the Y axis while the toolhead moves left and right on the gantry (the X axis). The Z axis is nothing more than a small gantry that moves the toolhead up and down, typically with a leadscrew.

The commands being passed to the controller on the CNC machine are called G-code. The commands passed to the controller (either via the onboard software or a computer passing G-code to the controller) move the toolhead to various locations in all three axes and runs the spindle motor so that the milling end can then cut into the material as desired. The same milling end will also drill all required holes to the desired depths. Of course, if the workpiece is to be separated from some stock material, it has to be secured to the worktable or surface so that it doesn't move while it is being worked on.

Some machines include a control console (a screen with some buttons or knobs, or a touchscreen) to allow operation as a standalone unit. However, the controllers cannot typically generate G-code, only a computer running suitable software can. This software reads a design file and translates it into motion commands that the controller can understand.

One of the most common motion control software for the CNC controller is **GRBL**. GRBL is open source firmware that can be installed on the controller that then receives the G-code from the computer or reads it from a file. Various vendors also have their own versions of GRBL tuned specifically to their controllers or machines, although most common desktop machines will run non-proprietary GRBL just fine.

GRBL is not the only option, however. **Marlin** and **Smoothieware** (most commonly used for 3D printing) can also be used, which allow 3D printer frames to function as CNC and even laser engravers. Some vendors even have proprietary machines that are 2-in-1 (3D printer and laser) or 3-in-1 (3D printer, laser, and CNC machine) that either have a second toolhead on the X gantry or support swappable toolheads depending on the desired operation. *Anycubic* produced a 2-in-1 3D printer and laser combination, and *SnapMaker* is famous for making a 3-in-1 unit.

Here is a picture of the microcontroller and touchscreen controller for Bumblebee:

Figure 1.8 – An MKS DLC32 controller with a TFT screen attached

The *MKS DLC32* board in the preceding figure is easily available on Amazon. The blue enclosure was 3D printed. The software (commonly called firmware) that configures the controller and touchscreen for CNC/laser has been loaded from sources made available by the manufacturer.

Ideally, the best case for a CNC machine is where repeatability and/or high precision are required at speed. CNC machines can make our object a lot more quickly than we can by hand, and the more complex the design the better the reason to use a CNC machine. Simple designs (for example, a rectangular panel with four holes) that only need to be made once are not an effective use of a CNC machine. However, parts that must fit together with high tolerance, perfectly rounded edges, and precisely rendered patterns and holes/fixtures are perfect uses for a CNC machine. For example, in *Chapter 5, Full CNC Workflow with Different Materials*, we will need to make several parts for a model airplane. Cutting several of these manually with a knife is tedious, but a CNC machine will make fast work of them, and they will all be cut exactly per the drawing we started with.

What is G-code?

G-code is nothing more than motion commands that the CNC machine's controller interprets to move the toolhead. Of course, those commands are passed on to the motor as the number of turns the motors' shafts have to make in one direction or the other. In addition to motion commands, there are commands to start and stop the spindle motor. All these commands are generated when software running on a computer interprets a design as a series of movements. There are many applications like this available, some free and others that can be purchased. Some design applications are also capable of generating G-code (for example, *Fusion 360*), which can then be passed on to the *sender* application, which in turn passes it on to the CNC controller. Some sender applications also generate G-code, such as *LightBurn* (for lasers), *Mach 3*, and *Easel*. For the purposes of our projects, we will focus on freely available firmware and sender software.

A note on laser machines

Since laser machines are mostly 2D machines (operating in X and Y only), G-code can be derived from simply an image (such as a JPG file) loaded into the sender software, which in turn converts it to G-code. Unlike in a spindle-based machine, the laser power can be controlled in G-code to also create shades so that various textures can be rendered on a surface.

G-code typically looks like this:

```
G01 X232.114456 Y13.456700 Z-1.5 F200.000000
```

G01 means move in a straight line to a specific position (here defined by the X, Y, and Z coordinates). F# defines the feed rate, or how fast the toolhead moves (in mm/minute).

As an additional note, I also want to tell you that while most 3D printers will run fine from G-code stored on removable media that is inserted into the machine's controller, not all commercially available controllers support this, and a PC will be needed to control the machine. This is not always the case. For example, appropriate firmware loaded on certain boards (such as the *MKS DLC32* and later boards), or boards that accept a touchscreen with its own removable media slot, allow the CNC machine to be entirely standalone.

Safety considerations for CNC and lasers

CNC machining involves the removal of material using a hardened metal bit that can fling fragments of your workpiece all over your shop. It should be needless to say that you should always have hand and eye protection on whenever you are working with and on your machine. This is especially true with lasers. The blue, green, or red laser on your machine can destroy your eyesight in an instant, so whenever the laser is on, you should have appropriate glasses on to filter out any harmful reflected laser light. Never operate a laser without those special glasses. Every laser unit I have ever purchased came with a suitable set of adjustable glasses (a green lens for a blue laser, for example). If your CNC bit shatters, you do not want metal fragments in your eyes or on your hands, and if your laser hits a reflective surface, you do not want to blind anyone looking over your shoulder or elsewhere in the shop with you. Do not allow pets anywhere near an active laser as they too could be blinded.

Summary

We now have learned some of the basics of CNC revolving around what it is, how it works, and what drives the design of various machines (for example, when to use belts instead of leadscrews). We also looked at how we might machine not just flat stock material, but also curved surfaces (with the rotary axis). All these are important concepts to grasp because they lay the groundwork we need to select, assemble, configure, and customize our own machines.

In the next chapter, we will look at setting up our own desktop CNC machine and get underway with fabricating some basic parts and shapes.

2

Setting Up and Configuring the 3018 CNC Machine

We have already covered some of the basics of **computer numerical control** (**CNC**), but in this chapter, we will focus our attention on a specific machine that is widely available from various vendors on multiple shopping platforms (*Amazon, AliExpress, eBay, etc.*). The 3018 CNC machine is a very basic, rugged unit, well suited to modification and upgrade, and its design has proven to be workable. You can choose to acquire a fully (or mostly) built unit, a kit to make one, or you can build one from scratch. I have tried all three approaches with similar results. However, before a purchase is made, some criteria and use cases should be established.

For this chapter, our objectives revolve around understanding the anatomy of a CNC machine, going through a primer on the selection of a suitable unit, and getting set up and running, as outlined here:

- Anatomy of a CNC machine
- Making the build-versus-buy decision
- Buying a pre-built unit
- Building your own unit
- Configuring, calibrating, and testing your CNC machine

Technical requirements

To operate your CNC machine, we should first discuss some basic requirements that will be needed to fully take advantage of your machine:

- **A Computer-Aided Design (CAD) application**: Unless you are downloading pre-generated G-code, you will need to either create or convert a drawing to G-code. There are several avenues for this. I use **Tinkercad** (a simple and free cloud-based CAD application) to import **STereoLithography (STL)** files or draw my own. From there, I save the drawing as an SVG file, which I subsequently convert to G-code using a conversion tool such as **JSCut** (`http://jscut.org/jscut.html`), which is also cloud-based. JSCut also allows me to visualize the G-code so that I can see how the cuts will proceed on the machine.

  TinkerCAD is not your only option; there are other more sophisticated packages. However, in my years of garage tinkering, I have had little need for more complex (and possibly more expensive) tools. You can create an account on Tinkercad at `https://tinkercad.com`.

- **A personal computer (PC)**: While there are solutions that cater to Linux-based or Apple/Mac computers, I have found that Windows-based software is the most prevalent and easiest to install and use. That is not to say a non-Windows machine is not viable, just that the offerings are fewer. The machine itself does not have to be a top-of-the-line machine. I run most of my CNC and laser machines using *Intel i5*-based machines with anywhere from 4-8 MB RAM and hard drives as small as 125 GB; most of the space is taken up by the operating system itself. The personal computer should have suitable USB ports (*USB 2.0* and later) as you will need these to operate most machines as well as upload firmware.

  You may also choose to run your machine using a **Raspberry Pi single-board computer (SBC)**, which you would use as both the machine controller and the G-code sender. For this, you will need to install a suitable daughterboard (often called a *hat*) such as the *Protoneer*. We will not be exploring this approach in this book, but the concepts should not be too difficult to extend to something such as this. Add a small monitor and a keyboard, and your CNC machine is also a fully functioning (albeit dedicated) PC. *Note*: You will only require the PC to generate G-code and *not* connect it to your CNC machine if you have another means to send G-code to the CNC controller (such as an LCD controller with an SD card slot).

- **Administrative rights to your PC**: You should have the ability to load drivers and install software on your computer. If you are using a shared or work computer, you will need administrative rights from your system administrator to be able to prepare the computer for use with the CNC machine, including unblocking USB ports (for security reasons, many corporate PCs have their USB ports disabled or severely limited).

- **An SD or Micro-SD card and suitable reader**: If this isn't built into your PC, SD card readers that plug into your USB port are easily available just about anywhere. I frequently use high-capacity Micro-SD cards that fit into an SD card shell, which is then inserted into a USB SD card reader. This allows me to move this storage media (the Micro-SD card) from PC to PC and from CNC machine to 3D printer to laser wherever an SD card slot is available.

- **Arduino IDE**: This is for compiling your own custom firmware. If you only plan to load precompiled firmware, you do not need this. Download the IDE from `https://arduino.cc`.

> **Note**
>
> As we deep dive into getting our machine set up and ready, you might consider heading over to `https://howtomechatronics.com/tutorials/how-to-setup-grbl-control-cnc-machine-with-arduino/`. This article breaks down CNC controllers further (many are Arduino-based) using a basic controller commonly available on *Amazon*. Look closely at the step calibration section because we will explore this deeper in this chapter as we get our machine ready.

- **GRBL .hex file**: You will either create this yourself (for a custom implementation) or download a specific vendor's implementation of **G-code Reference Block Library** (**GRBL**). You can also download the baseline version from the GRBL GitHub repository at `https://github.com/grbl/grbl`.

- **z-axis setting probe**: While optional, I prefer to use this little tool (available from several vendors) to ensure I set the origin point (*Z=0*) for my z axis. The tool is cheap and precisely sets the origin point on the workpiece no matter what its thickness. If you have endstops on your machine, you may not need this as much because it is a simple calculation to determine where *Z=0* as the top of the z-axis limit is a known value and is the distance from the tip of the carving/cutting bit to the worktable. The thickness of the wasteboard and workpiece is added and then subtracted from this height to determine where *Z=0* is. I much prefer to let the machine determine this instead, which is why I used this probe, especially since I may have different-length bits and different-thickness materials. It just gets tedious to keep measuring the differences, so of course, I resort to automation. There is a great tutorial on how to use the probe at `https://buildyourcnc.com/PrimeronHomingandLimitSwitches.aspx`.

Anatomy of a CNC machine

Some of the basic components of a CNC machine were touched on in *Chapter 1*, *The What and Why of CNC*, but ahead of deciding which machine suits our purposes, we should know what the particulars are of each component. Here's a view of a 3018 machine I built using some left-over frame parts from other projects and some 3D-printed components. The overall part count for the frame is very low, and most components can be obtained off the shelf. Most 3018 machines will look a lot like this:

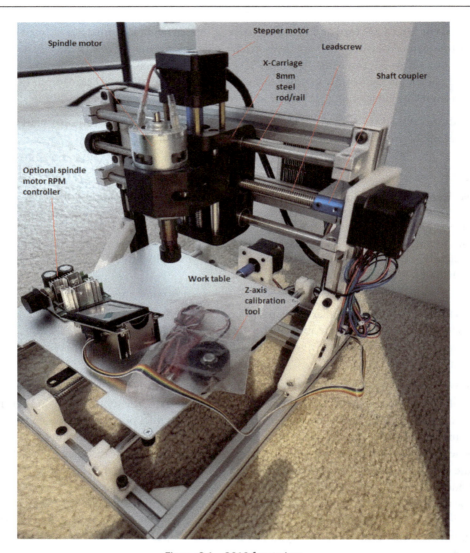

Figure 2.1 – 3018 front view

Looking at the labeled components in the preceding photo, we have the following:

- **Spindle motor**: We mentioned this component in *Chapter 1, The What and Why of CNC*. This is part of the toolhead and what drives the endmill bit. Think of it as nothing more than the motor out of a common hand drill with a chuck at the end of the drive shaft that will accept any of a variety of endmill bits. A review of commercially available machines may refer to this as a *775 Direct Current (DC) brushed motor*. This is what many kits and pre-built machines come with. This particular motor came with a chuck already press-fit onto the motor's drive shaft.

- **Motor RPM controller**: This optional item allows the user to externally control the **revolutions per minute** (**RPM**) of the spindle motor. Some older versions of GRBL can turn the spindle motor on/off and reverse its direction. After version 0.9, it became possible to use **pulse-width modulation** (**PWM**), a series of pulses turning the motor on and off and varying the duty cycle (how long the motor is running) to control the motor speed. In the absence of PWM, and when an upper limit for RPM is desired, the motor RPM controller comes in handy.

- **Stepper motor**: This motor is characterized by the ability to precisely control rotation, including holding the motor shaft stationary. These motors come in multiple designations (for example, *NEMA 8, 14, 17, 23*, and *34*, which identify the size of the motor). Additionally, these motors have specific *holding torque* (the amount of force the shaft would need to experience to move it out of position) properties. Our 3018 machine uses three such motors, one for each axis (x, y, and z).

- **Leadscrew**: This is a helical precision screw that the stepper motor rotates to achieve motion in the x, y, or z axis. These also come in a variety of sizes (diameters and pitch), but for our purposes, we are using *TR8* (8 mm diameter) leadscrews, which are very common. These have differing pitches (how much motion the leadscrew nut travels through each revolution of the leadscrew). For example, a 2 mm pitch means that the nut will move 2 mm with each rotation of the leadscrew. It should be clear that very precise movement can be made with these components.

- **Shaft coupler**: This connects the motor shaft to the leadscrew. Since the motor shaft is typically 5 mm in diameter and the TR8 is 8 mm, the coupler will have one end with a 5 mm bore and the other with an 8 mm bore. The coupler is secured to the leadscrew and motor shaft by tightening the screws on the side for a friction fit. A common problem with kit-built machines is having loose couplers, which interfere with the proper motion of the toolhead. Generally, couplers should be inspected after the build periodically to ensure they haven't loosened.

- **X-carriage**: This carries the toolhead and the z-axis assembly. Unlike 3D printers, where the z axis is typically independent (*CoreXY*, *D-Bot* machines) or moves the x-axis gantry, many CNC machines carry the entire z-axis mechanism and move it left and right (in the x axis). This means this is a very heavy component, including the rails, z-axis frame, z-axis motor and leadscrew, and of course, the toolhead itself. z-axis travel is typically small and is represented by a negative number in G-code (where Z=0 is the surface of the material being machined).

- **8 mm steel rod**: Not all CNC machines use these smooth, very durable rods for rails. However, many 3018 machines you will encounter use this as a standard for their motion systems. Other alternatives are linear rails, which are smoother, exhibit less wear over time, and are even more accurate (straight). They do, however, tend to be more expensive.

- **Worktable**: This is the surface on which you will place your materials to be machined. For my DIY machine, I used parts I already had in my parts bin, which included the metal platform from a long-defunct 3D printer. Many of the machines you will find online have extrusions in place of a worktable because those have slots that allow you to fasten clamps or even a *wasteboard* to the worktable (see *Figure 1.1* in *Chapter 1, The What and Why of CNC* for another of my machines with an extrusion for a worktable). As the machine in *Figure 2.1* was built explicitly for this chapter, we will later show some examples of how to secure workpieces of a wasteboard to the table.

- **z-axis calibration tool**: Because my machine is not using endstops (more on this later), we need to, at a minimum, determine where *Z=0* every time we start a job (unless every piece of material we start off with is identical to the one before it). By setting *Z=0* with this tool, we can be sure that our cuts (especially where we need to cut to specific depths, such as when drilling holes) are very accurate. This tool is not attached to a controller all the time, but rather is removed once the *z* axis is *homed* (that is, set to its lowest limit; in this case, Z=0).

> **A note on wasteboards**
>
> Wasteboards are consumables where CNC is concerned. Ideally, you would never need these because you will have precisely homed your *z* axis and your material to lie perfectly flat against the worktable, and your design ensures that your endmill will never touch the worktable. Of course, that implies a huge amount of trust in a machine, and no machine can be 100% trusted to perform exactly right all the time, every time. A wasteboard ensures that if there is an error, any damage is confined to it.

Now, let's have a look at another view of the same machine. *Figure 2.2* gives you a closer look at the same components we discussed up till now in this section, and we can see a few more:

Figure 2.2 – 3018 top view

Notice how the screws are holding the worktable to the frame underneath (which was harvested from a dead 3D printer) and are countersunk into the worktable. This ensures that whatever is placed on top of the worktable lies flat and can consume all the real estate presented by it. Looking at this picture, you see the 775-motor mount. There are different-sized motors that dictate what you have: a mill or a machine. The same motor mount (if you look very, very closely, toward the front) has these small notches/grooves cut into the inside of the mount. These allow you to insert a laser toolhead (which is typically square-shaped) into the same mount whenever you want to switch from CNC to laser and back. On one of my other machines, I have fastened my laser head to the side of the motor mount so that I don't have to keep swapping toolheads.

> **3D printer and printed parts for this machine**
>
> This machine makes extensive use of spare parts in my parts bin as well as 3D printed parts found on `https://www.thingiverse.com`. Most designs on this site are free to use for non-commercial purposes without prior permission. There are other sites where you can download models and designs for a fee (such as `https://cults3d.com/en`), but there is plenty to explore at *Thingiverse* and *Thangs* (`https://thangs.com/`). These two sites are where I usually go for part designs when I am building from scratch. Most if not all the parts I 3D printed can also be purchased at various online shopping sites such as *Amazon*. In my case, I purchased an X-carriage (which includes the *z*-axis assembly) because the 3D-printed equivalents proved to be unreliable for heavy-duty use on previous builds.

Looking at *Figure 2.2*, we can see the following components:

- **LCD controller**: This component is entirely optional, and allows for the operation of the CNC machine as a standalone unit. The computer is not needed to control the machine, only to generate the G-code, which can be copied onto an SD card and loaded into the controller.

- **Motor mount**: Stepper motors are secured here. In *Figure 2.2*, you see two of these for *x* and *y*, while the *z*-axis motor mount is largely made of brass standoffs. Because of the orientation of the motor, different forces act on the motor. In *x* and *y*, these are largely the same, but in *z*, you have more compression/tension loads owing to gravity and the need to pull and push a load with the leadscrew against or with gravity.

- **Endmill chuck**: This is just the part in which the bit is inserted. The bit is secured the same as you might in a hand drill. You hand tighten the nut at the bottom of the chuck, and that secures the bit in place (unlike a typical drill bit, however; endmill shanks are square to reduce the likelihood of the bit slipping in the chuck).

- **x-axis bearing**: On the 3018 there are two similar bearings. One of them is this one, for the *x* axis. While you could have a setup where this bearing does not exist, your *x*-axis leadscrew will wobble, which will lead to inaccuracies in movement. On desktop 3D printers, many manufacturers leave the *z*-axis leadscrews free on the non-motor end, and you would be well served to add a bearing there to prevent wobble. The other place you have a bearing such as this is on the *y* axis, which moves the worktable forward and backward.

Finally, let's have a look at the rear of the CNC machine in *Figure 2.3*:

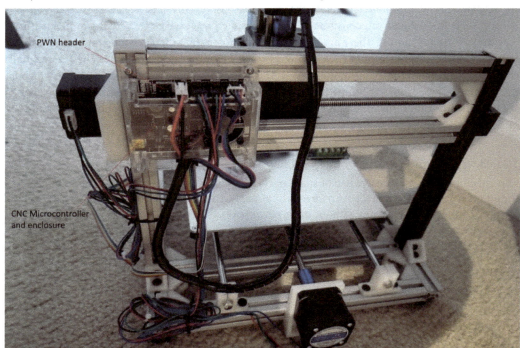

Figure 2.3 – 3018 rear view

In the preceding photo, you can again see the same familiar components as in *Figure 2.2* and *Figure 2.1*. Notice the 8 mm rod mounts here. When I was building this machine, I ran out of the aluminum ones and I didn't want to buy more (and wait for them to arrive), so I found a design online (and I did one of my own) and 3D printed it. Again, these are parts you can get online, no 3D printer required; I was just impatient to get moving. I printed eight of these in a few short hours.

In *Figure 2.3*, you can see the microcontroller mounted on the left. This unit has a USB port for connection to a computer so that G-code can be passed to it directly line by line. However, also of interest is the PWM header at the top left. If we had a motor that we could control that way, then we could use PWM to control motor speed. However, my criteria included laser support, and I wanted my laser to be controlled by PWM (so that the laser pulses instead of having a constant beam). The PWM header is therefore set aside for the laser toolhead I am planning for this machine later. The controller is encased in an enclosure with a fan over the central processor, and all electronics are powered by a laptop-style power supply.

Making the build-versus-buy decision

Now, some of you may be wondering whether to buy or build your own CNC machine. There are pros and cons to both options. I chose to build my machine because I primarily wanted to use spare parts I had from other builds that I knew might require some customization. I also wanted some freedom to add functionality that might be limited by an off-the-shelf kit.

However, based on what you want to do with your machine and how much time and money you have to spend, you can also choose to buy a complete machine.

To make the buy-or-make decision easier for you, in this section, I will highlight all the considerations that you may need to take into account, based on my personal experience.

Before we begin, I want you to ask yourself the following questions:

- What is the overall budget for the machine? Consider the cost of building as well, and factor in the amount of time you will spend in assembly.

- Which materials will be milled/machined?

- How thick will any given workpiece be?

- How large will the workpieces be? Is a typical 3018 work area enough? Is a desktop-sized unit sufficient, or would a *workshop*-sized unit be more suitable?

- Will a laser be added afterward? Or is a laser what is primarily desired? Will there be a need to work on curved surfaces?

- What size/power motors are available for the proposed budget? Are they sufficient for the workload anticipated?

- What kind of controller is desired? Is a commonly available purpose-made CNC controller such as the one in *Figure 2.3* sufficient, or is more sophistication wanted, such as an MKS DLC board, mentioned under the *How CNC works and when to use it* section of *Chapter 1, The What and Why of CNC*? Is a touchscreen controller desired?

- What sort of features are desired/required? Are endstops important? What about a dust vacuum attachment? Can features such as that be added later? Are the added features widely available off the shelf or will you have to fabricate them yourself (for example, for my 3018, I will have to make my own mounts for the endstop switches)?

- Will it be possible to move the workpiece around, or is it more practical for all the axes to move around the workpiece?

The last question asked can dictate what sort of machine you choose. For example, *BumbleBee* (*Figure 1.2* in *Chapter 1, The What and Why of CNC*) was designed to operate on workpieces that are too large to be moved by *NEMA 17* motors, and my design criterion was to minimize the use of heavy-duty motors. Consequently, Bumblebee sits on top of a workpiece, and the toolhead moves in all axes around it. This allows for the positioning of the whole machine over a specific piece of material that can be as large as I want it to be and then cut from it directly. On the 3018, you would have to cut your raw stock to fit on the machine first.

When you select a machine, consider carefully what your use cases are before you invest time and money into a specific unit. Still, even though we are focusing on the 3018 here, all the concepts apply to most other CNC and laser machines.

Buying a pre-built unit

Once a selection is made, don't rush off and buy it just yet from the first vendor you see (for a pre-built unit). Instead:

- Comparison shop and be open to giving up a nice-to-have feature (such as a dust vacuum head) out of the box in favor of a substantially lower cost.
- Consider the frame materials used; not all 3018 machines are alike. Some use aluminum extrusions, some use thick acrylic panels, and others (the more expensive ones) use aluminum panels for the gantry. Considering the materials you will be milling, is a less rigid machine (with acrylic or melamine parts) acceptable? What happens if the acrylic cracks? Will you be able to replace it?

Building your own unit

If you are thinking of building your own unit, you must be even more selective. The following are some example considerations:

- What is the recommended holding torque for your motors?
- Can you locate the right leadscrews with the right pitch? Are you going to need a Delrin (a type of low-friction, high-wear-resistance plastic) leadscrew nut or a brass one? *BumbleBee* uses Delrin nuts, but my 3018s all use brass nuts. The difference is often dictated by design and expected wear.
- What power/RPM spindle will you use? Is a plastic mount sufficiently rigid, or will you need a metal one?
- If you are building a desktop unit, you will likely have a worktable that moves. If you are building a unit with a larger work area, the entire gantry will have to move along y. Are you going to use belts, and is that motion system likely to have binding problems (I have seen some of the belt-driven systems get hung up because the belt gets trapped in the frame extrusion slots and wears down very quickly, or loses tension very easily)? Look at the motion system closely and play the devil's advocate to see where the potential for wear is and whether you will be servicing the machine more often than you would like.

Regardless of whether you are building or buying your CNC machine, you will have to go through some setup processes before confidently working on a project.

Configuring, calibrating, and testing your CNC machine

Once you have fully assembled your machine, you will need to set up the parameters that control it, prepare it for calibration, and run some tests before you put it into operation.

If you built your own machine from an open source design, you might also have downloaded a copy of the firmware for your microcontroller. In many cases, the controller already has firmware on board, but it may be obsolete; so, the first thing to do is determine whether you have the latest version installed. If you purchased a pre-built unit, then check with the vendor to make sure the firmware is up to date.

For off-the-shelf units/kits that come with everything, do not immediately overwrite the existing firmware with a generic version unless that is what is there already, since some vendors create their own flavor. Check with the vendor first to ensure you have an up-to-date installation, or, if you want to install your own, get whatever configurations they may have added and save yourself the headache of having to figure them out on your own. A great source of information is *Endurance Lasers*, which publishes a bunch of useful articles on CNC and lasers. For GRBL command information, you can go to `https://endurancelasers.com/an-important-things-you-need-to-know-about-grbl-firmware/`. You can also download the latest version of GRBL from its GitHub repo at `https://github.com/gnea/grbl`.

A machine built from scratch is going to require you to set up things such as the spindle motor thresholds, the step measurements (how much movement you get from each turn of your leadscrew), the limits of your work area for your machine, and much more. There is even a link to a firmware image just for 3018 machines. For some of my self-designed machines, I started with default settings, attempted to cut a rectangle of known size, and then measured what the actual rectangle dimensions were and adjusted the numbers until the rectangle was cut to match the dimensions of the design in the sender program.

Step calibration

Let's touch on how to determine the settings for your leadscrews (there is a similar calculation for belts, but since the 3018 uses leadscrews, we will use this). Obviously, if you are buying a kit, you will not have to do this. As mentioned before, your controller may have firmware already loaded that has some default value for x, y, and z. Your machine may be set up for imperial or metric measures. The vast majority of machines I have encountered have been metric, so the example we will go through in this section will be using calculations based on the metric system. Let us first gather some information that we know about our motor:

- **Number of steps per revolution**: This is defined in the specification of the motor and should be published in a pamphlet that comes with it. Let's go with *200* steps per revolution for our example.

- **Number of microsteps**: These are defined by the stepper driver on your controller and indicate the number of microsteps per full step. This number should be published with the specification of the driver for your controller board. If you cannot find it, you can assume a number, run the calculation, do a test, check the outcome, and then adjust this number accordingly. Some of the offline/LCD controllers allow you to adjust this value for each axis. For the sake of argument, let's use *4* for the number of microsteps. This means that each revolution of our motor will require *200 x 4* (or *800*) microsteps per revolution of the motor.

- **Leadscrew pitch**: The common TR8 leadscrews come in varying pitches (measured in mm) typically denoted with a *p*. For example, *TR8*8-2p* means an *8 mm* diameter lead with a *2 mm* pitch. The pitch value indicates the number of millimeters the nut travels per revolution.

So now, *200* steps per revolution translates to *2 mm* of travel per revolution, so we have *200/2* steps per mm or *100 steps* per mm. This also translates to *400* microsteps per mm (*800/2*).

If you are compiling the firmware, you would be setting the values in the firmware code (for the *z* motor, for *x* and *y*, the same setting is defined by the axis) as follows:

```
#define DEFAULT_Z_STEPS_PER_MM 100.0
```

The GRBL command to determine what the current value is for *z* is $102. For *y*, the command is $101, and for *x*, it is $100. You can enter these in the sender program once you are connected to the machine to find out what the current firmware settings are.

If you don't want to download and compile the source, you can enter the appropriate values in the terminal window of your sender software as follows:

```
$102 = 100
```

Now that we have determined the leadscrew settings, let's look at how to load precompiled firmware and then make changes to the settings for our machine.

In the article from *HowToMechatronics* mentioned in the *Technical requirements* section, calibration is described in terms of using the wizard provided by the **Universal G-code Sender (UGS)** (see the *G-code sender software* section later in this chapter for an evaluated list of G-code sender applications) to set your steps. The onboard wizard starts with the default of *250* steps per mm and allows you to adjust this based on the amount of movement you see; for this, we will need a metric ruler with millimeter measurements. The process is fairly simple and involves the following steps:

1. Start UGS and connect to your machine.
2. Position the toolhead somewhere in the work area and make a note of that position. If you have an endmill bit in place, lower the toolhead with the *z* axis (using UGS) so that the endmill touches the work surface (use a piece of scrap wood on your worktable). Mark this location and raise the toolhead a few millimeters.
3. Now, move your machine along the *x* axis using the **X+** button. The toolhead should move from left to right *250* steps. Lower the toolhead again and mark this location.
4. Measure the distance between the two marked locations. In the wizard, enter this distance. The software will then calculate the correct steps per mm for your machine for the *x* axis.

5. Repeat the same process for *y*, pressing the **Y+** button to advance in that direction.

6. For the *z* axis, you will need to measure the height you move the toolhead with each press of the **Z-** button (note, not **Z+** because raising the toolhead is considered negative in CNC since the workpiece surface is Z=0). Here, you simply measure the height from the worktable to the tip of your bit and again enter that value into the wizard.

Once you have entered those values and saved them into the firmware, your machine is largely ready (other than homing, which I will address in a later section when we begin testing and make our first cut).

Firmware flashing software

To get your version of GRBL on your CNC machine's controller, you will need specific software. Loading firmware is called *flashing* and makes it possible to use a variety of tools. If you are building a machine from scratch, I encourage you to select a board that already has a bootloader (effectively a program in memory on the controller that allows you to load updates to the firmware). Without a bootloader, flashing firmware is a little more complex.

While there are many ways and tools to flash firmware, the following are two popular methods:

- **Using XLoader**: This is the most common method to load firmware but is only available for Windows. The downside of this little program is that it *could* hang up, wiping out the firmware on your board in the process. I have had it work fine on some PCs (*Windows 10*) and not on others. Frequently, the issue has to do with the communications speed (*baud rate*) at which your USB port can send data to your board. You will have to experiment here or ensure you know what the numbers are for your controller. You can download XLoader from GitHub at `https://github.com/xinabox/xloader`. Otherwise, the only thing you will need is the `XLoader.exe` file in the ZIP file you download.

The following is a screenshot of the XLoader interface:

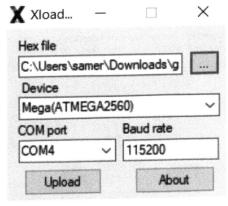

Figure 2.4 – XLoader interface

- **Using LaserGRBL**: This is not only a sender program but also has a built-in process for flashing firmware. I have used this program to flash firmware on both CNC and laser machines, adjust the firmware (stored permanently onboard the controller), and control some of my machines. Here is the main screen of the latest version I have. Note the menu option where GRBL can be uploaded:

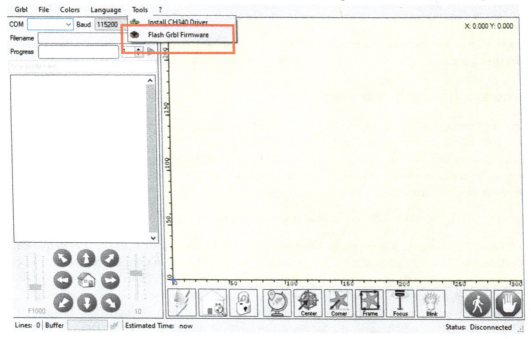

Figure 2.5 – Main screen for LaserGRBL

Once you select the **Flash Grbl Firmware** menu, you will be led to a dialog box that is very similar to XLoader's:

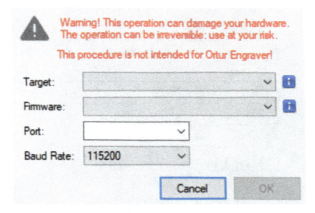

Figure 2.6 – GRBL loader dialog box in LaserGRBL

There are other ways, of course, including freely available versions of *Benbox* (another laser engraving program that allows for custom versions of GRBL to be loaded from the Benbox vendor), but I have found the two methods I have discussed easy to use. Also worthy of mention is *T2Laser*, which also allows you to flash versions of GRBL from a menu of available distributions. However, your mileage may vary here, especially if you have your own custom-compiled version.

Your precompiled GRBL firmware file will have a `.hex` extension. This is what you will need to provide to the firmware loader. You will also need to define the target COM port, represented by your USB port. I have noticed some loaders are not too comfortable reading the firmware file from network shares, so it is recommended to have the file loaded on the PC's local drive somewhere.

> **Connecting a controller to a PC**
>
> Your CNC machine will appear as a USB device on your PC. I have had excellent success connecting various desktops and laptops to many devices and controllers with little to no issues. Frequently, the device driver is automatically loaded (depending on the nature of the controller). However, failing that, you may need to load the device driver, commonly referred to as the *CH340* driver. The reason for the name is that this chip (the *CH340*) is a common component of Arduino boards on which most hobbyist controllers are based or are clones. Load this driver (it should come on storage media with your controller) if you are unable to get your PC to recognize the board. When I am working on a from-scratch build, I check connectivity and load drivers and firmware before the board is ever connected to the CNC machine itself. Once connected, a look into **Device Manager** (for *Windows*) will show you which port your controller is on (*COM4, COM5, COM6*, and so on).

It is important to pay attention to the baud rate once you are connected to your controller. Using an incorrect baud rate will cause some loaders to hang, and you will have to force close them and try again. The acceptable communication speed should be published by the vendor of the controller in their specifications.

Before we move on to looking at G-code sender programs, let's briefly go over LCD controllers. There are many alternatives here, and how you connect depends on the boards you are using. If you are using a controller such as a Mana board, there is no provision to use an LCD controller, and you are limited to using a dedicated PC for the duration of your CNC operation. On the other hand, more modern controllers will have connectors, such as those you saw in *Figure 1.6* in *Chapter 1, The What and Why of CNC* to add an LCD.

I am a big fan of the LCD controllers put out by *MKS Base* (or *Makerbase*). These come in a variety of sizes, with support for SD or microSD cards so that the LCD controller can stream G-code to your machine instead of your PC. Many of MKS's products also include the provision for Wi-Fi to allow wireless control. Here is a photo of a *2.4"* touchscreen unit I have set aside for my 3018. These units come in various other sizes, including *3.2"* and *3.5"*. When you purchase them, they may have the 3D printer firmware on them by default. However, MKS puts out CNC and laser firmware as well.

In *Figure 2.7*, you see that I have already loaded the CNC/laser firmware:

Figure 2.7 – The MKS TFT24 for my DIY 3018

The following photo shows you my *FLSun Cube printer*. Same LCD vendor, but a different model. The 3D printing firmware is FLSun's modification of the baseline provided by MKS:

Figure 2.8 – Another TFT LCD with the default 3D printer firmware installed

Loading the firmware is very straightforward and requires the LCD to be connected to your controller and the controller connected to power. Prior to starting up, you load the desired firmware on an SD card and insert that into the LCD controller. When you connect power and the devices boot, the LCD automatically recognizes the firmware file and loads it. The great thing about this is that the LCD firmware will have most of the features to support upgrades, such as adding a laser, supporting endstops, and so on.

G-code sender software

Let's review the various G-code sender programs and the ways of loading firmware onto your machine's controller. Recall that we need senders to send the G-code to the controller because most small microcontrollers have limited memory capacity to hold more than a small number of commands. The other alternative is to use an LCD controller that allows offline operation, but you still will need sender-type software to generate the G-code file to load into the LCD controller, especially given the availability of calibration wizards such as the one in *UGS*. In this brief compendium, I list some of the more popular applications with as much focus on multiplatform availability as possible:

> **Alternatives to GRBL**
>
> While we are focusing our attention on GRBL in this book, you should also be aware of other firmware that is available, such as *Maslow CNC*, *TinyG*, *Marlin*, *Repetier*, and *Klipper*. For example, while Marlin is very common for 3D printers, it can be modified/adapted to run a spindle instead of a hotend (the part of the 3D printer that lays down melted plastic).

- **UGS**: This is one of the most common, easiest to use, and feature-rich applications. It is 100% free and is multiplatform, covering *Windows*, *Linux*, and the *macOS* world. It does support Raspberry Pi as the host platform as well, which means your CNC machine can have an embedded SBC PC, or simply be connected to one and make use of your machine's onboard CNC controller. The other benefit is that you can also use its wizard to calibrate your machine (as discussed previously) and the ability to check the wiring of your motors. You can also configure your endstops or set soft limits (no physical endstops). Download it from `https://winder.github.io/ugs_website/`.

- **OpenBuilds CONTROL**: I like this fairly light application because it has the basic G-code sending functions but also includes the ability to flash firmware. As new versions of GRBL become available, you can update your machine using this software. If your PC is also on a network, you can also remote control your machine over the same network. This particular application has a huge following and can be used to control a CNC machine, a laser cutter/engraver, a drag knife, and a plasma cutter. Support for Windows, Linux, and macOS is available. Download it from `https://software.openbuilds.com/`.

- **ChiliPeppr**: If you prefer a browser-based application and can have a persistent web connection throughout your CNC machine's run, then consider this application. ChiliPeppr not only is multiplatform but also supports multiple firmware platforms. You do need to install a small serial port JSON server to allow the application to talk to your CNC machine, but that makes for a very small footprint. Needless to say, you can use a Raspberry Pi, Linux, Mac, or Windows PC to run the machine. It does have limitations in that it can only handle *25,000* lines of G-code in its buffer. You can reach the application and download the JSON server from `http://chilipeppr.com/` and go to the GRBL workspace at `http://chilipeppr.com/grbl`.

Other available applications include *Candle*, *CNCjs*, *Easel* (subscription-based), *Ultimate CNC*, and *LinuxCNC*. Consider these as well if you would like to explore alternatives to what has been presented previously.

In the laser world, I have had good experience with three applications. The first is *T2 Laser* (`https://t2laser.wordpress.com/`), which I have used to flash firmware on both laser and CNC machines and actually can be used to control both. T2 is paid software, and if you select it, you will need to decide if you are going to use it on more than one PC because the license can be tied to a PC. If you plan to use multiple machines, purchase the license that comes with a dongle that allows you to use different machines (albeit one at a time).

By far, *LightBurn* (`https://lightburnsoftware.com/`) has been my favorite (albeit paid) application for laser work. Finally, *LaserGRBL* (`https://lasergrbl.com/`) has proven to be a very robust application for my purposes.

You may also run across a software called *Benbox*. While commonly available with some laser machines, some units come with modified versions of GRBL specific to the board on the unit. I have occasionally resorted to Benbox to test my laser and flash GRBL but I generally don't use it because of its limited features.

Running your first test cut

With your machine now set up and operational, it's time to cut your first piece of material. Before we begin, we will need a *wasteboard*. This is a piece of material that we can make mistakes with that will protect the worktable from the cutting/carving bit should we need to cut through our workpiece. Make this out of a piece of any plywood or MDF that you can get at your local big-box DIY store. I like these to be at least *1/4" (6 mm)* thick to ensure I have some leeway to stop a cut if it gets deeper than planned and to protect my metal worktable, and to drill holes to screw in clamps to hold my workpiece.

If you prefer, you can also buy wasteboards from places such as *Amazon*, but those are typically configured for worktables made of *80/20* extrusions, such as my machine in *Figure 1.1* in *Chapter 1, The What and Why of CNC*. Those have countersunk holes to secure the wasteboard using T-Nuts. If you have a flat plate, as I do with my machine in the photos in this chapter, use strong binder clips to hold the wasteboard in place.

Place your workpiece on top of the wasteboard. You can secure it to the wasteboard in one of two ways:

- *Option 1*: Drill holes in the workpiece through and into the wasteboard and screw it onto the wasteboard directly. Make sure that your screws are countersunk so that they don't interfere with the movement of your toolhead and that they are located outside the outermost boundaries of the object you will *liberate* from the workpiece. You will also need to drill a hole and put a screw through in a suitable location so that the part you cut out does not move around as it gets cut out of the workpiece. If you are only engraving to a specific depth, you don't need to do this.

- *Option 2*: Use clamps that your machine may have come with to secure your workpiece. Again, place those clamps where the toolhead is unlikely to collide with them as it carves through your material. Once again, if your final product is meant to separate from the stock workpiece, you must secure it to the wasteboard so that it doesn't come loose and mess up your cut.

The following is an example of one of those wasteboards you can buy on one of my 3018 machines. Note all the holes in it that are countersunk with T-Nuts as well as the holes for securing it to the worktable:

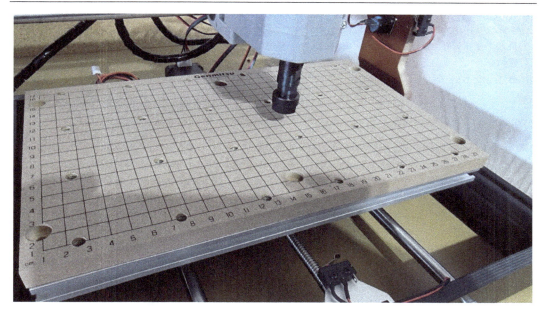

Figure 2.9 – A wasteboard on one of my 3018 machines

The final step before we make our first cut is to home our axes. Let's start with the z axis. The following photo shows one of these probes that I use. The way the probe works is to complete a circuit when the carving/cutting bit's tip touches the probe:

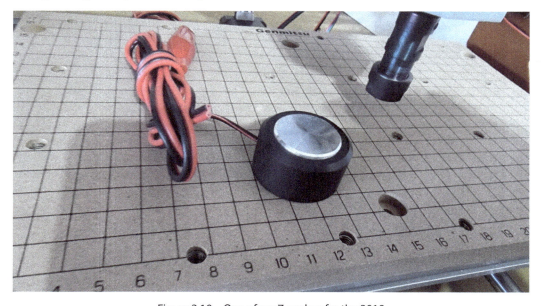

Figure 2.10 – One of my Z-probes for the 3018

The Z-probe typically comes with instructions, but fundamentally, you are connecting it to the controller (or if you are using an Arduino, to the *A5* and *GND* pins) and probe header, and then you set UGS to set *Z=0*. There is a great video on *YouTube* that illustrates this far better than anything written: `https://www.youtube.com/watch?v=PtJF8q3RrDo`.

Next, set the *X=0* and *Y=0* points on your workpiece and, using UGS in the **Machine Control** menu, jog your axes to where you want *X=0* and *Y=0* to be and press the appropriate **Reset Zero** button.

This *YouTube* video also offers some additional detail on getting your axes ready: `https://www.youtube.com/watch?v=A1z1L3q23HI`. The presenter here uses a technique similar to how bed leveling is done on a 3D printer where Z=0 is set using a piece of paper. Something to keep in mind is that this is something you can't really do as easily with some LCD controller firmware, which is why if I build a machine to be standalone, with an LCD controller, my preference is to have endstops to ensure the toolhead does not crash into the axes limits and I can home (set the origin) automatically every time. My DIY machine started out with no endstops but did end up with them just because it was far easier to calibrate.

With your axes set, it is time to run a test. I suggest you attempt to cut some basic shapes first: a triangle, a rectangle, and a circle. Using whatever design software you have, make these shapes, and set their depth to be smaller than the thickness of your workpiece so that you can just engrave the shapes. The easiest way to do this is to create the shape in your favorite CAD application and convert it to G-code. Load the shape in, generate the G-code, and then load the G-code into UGS (if that is what you are using). Start the spindle, execute the cut, and see the results. Once the machine is done, jog it back to its 0, 0, 0 location, stop the spindle if it is still turning, and turn off the machine so that you can inspect your results. Confirm the dimensions of your shape against what you intended them to be, and if everything measures up, you are ready to start doing some CNC milling.

Summary

In this chapter, we took a deep dive into what is needed to set up our CNC machine and prepare it to mill our workpiece. We also determined how to load firmware, calibrate it, and set the origin of the axes. Finally, we also learned how to prepare the necessary G-code to load into our machine. While we focused on a machine without an LCD controller, the only difference between using a sender program such as UGS and the LCD is that the LCD will require the G-code to be loaded onto an SD card and read directly from there. There would be limited functionality to reset zero on the axes, which then speaks to having endstops on your machine, something we will cover in greater depth when we get to upgrades.

Our next chapter will get into selecting our materials for milling, as well as the bits we need to use. We will also look into what and how various bits cut and when we should use them.

3
Understanding Material Properties before Making the First Cut

The CNC machines we work with at the desktop level have limitations both in terms of how big a piece of raw material they can process and what the material itself can be made of. Every material has specific properties that require attention when being milled. These properties include hazards from cuttings/dust, viable milling speeds, and heat dissipation, among others.

By the end of this chapter, you will understand the following:

- The fundamental properties of certain materials to help inform the process of material selection in your projects
- How to tailor your cutting methods, settings, and preferences to your specific materials, machine and parts you are fabricating

You will also be able to select materials that are viable for your specific project (for example, a generic suggestion to use wood is not enough since you may need material that is light, or durable, or flexible). An important benefit is that you will understand the hazards of using specific cutting methods with certain materials. Finally, as part of your learning, you will better understand how to select cutting bits for your CNC machine based on the properties of the material you intend to machine, and will be able to set up the types of cuts that are appropriate for your material and project.

We will address these properties by type of material and also discuss what sort of end mills and bits you will need for cutting the following:

- Hardwoods, plywood, and balsa
- Foam and composites (these materials can also be cut with a hot wire, but that use case is different and is covered in *Chapter 9, Project: Building Your Own 4th Axis*)
- Plastics and PVC
- Aluminum and other soft metals
- Selection of an appropriate bit or end mill

Cutting hardwoods, plywood, and balsa

Wood is the easiest material to mill with desktop machines. The material is relatively plentiful and much of what we do with wood can also be applied to *engineered* wood such as MDF, pressed/particle board, and manufactured board. I have recycled materials from scrapped IKEA furniture for years and cut it with a CNC machine and laser for all sorts of projects. You can even make some interesting engraved art with the laminated stuff. A lot of what I do extensively recycles materials like this to minimize waste (I recently built a full flight-simulator panel using engineered wood repurposed from an old foosball table).

Softer woods, such as balsa, can't be cut effectively with a high-speed end mill if they're too thin. The fibers will quickly shred and destroy the workpiece. This is also true for any wood, but really soft wood is most susceptible. If I am cutting thin bulkheads for a model airplane it is far more efficient to use the laser on my 3018 rather than the spindle. Hardwood, such as maple, should be cut with multiple passes using a sharp bit. Do not skimp on your cutting tools when working with hardwood. Anything thinner than 3 mm (⅛") I usually relegate to the laser just because the cuts are cleaner, especially with softer materials (e.g., light plywood, balsa, or some foam boards). While MDF and plywood can be cut with a laser, be careful with those as the resins/glues in them can burn if exposed to a powerful beam moving slowly over the material. I also generally do not cut small parts out of MDF and plywood because the material can shred when making lots of turns over a small area.

It is always a good idea to try various settings on a piece of scrap. Attempt to cut basic and complex shapes at various travel speeds and spindle RPMs. Some woods you can **upcut** and others you should only **downcut**.

> **Note**
>
> How you cut into material is defined by whether you are cutting through a piece of material or not.
>
> A *downcut* preserves the surface and typically preserves the surface of the material, preventing fraying and shredding. This is especially good for engraving and a suitable *downcut* end-mill bit is what you would need, but note that chips and cut material do not flow upwards out of the hole or opening created.
>
> An *upcut* preserves the bottom of the material and is meant for cutting through the material. I use upcut bits to cut parts that need to be separated from the stock material and to carve edges. Chips and cut material are pulled up with the flutes of the end mill, which may result in some fraying that can be remedied by some light sanding afterward.
>
> A *compression* bit combines both an upcut and downcut bit so that you can have clean surfaces on both sides. However, it is not a panacea for all projects since the first pass must happen below the upcut portion of the bit. Otherwise, the bit will behave like any other upcut bit. I like to use compression bits when I have surfaces that are soft and prone to shredding, such as MDF and plywood.

When selecting your material, consider the possibility that you might need to do more than just cut out a component, but also shape it with holes, grooves, joints, shelves, and similar artifacts. For that you might feel you need to use multiple bits on the same workpiece. This is another reason for your 3018 to have endstops so that you can home it and so that the machine always has the same point of reference from bit change to bit change. You might first drill all the holes, then cut out any interior holes and then cut the whole part out of the stock using two or three bits in the process. Most of our projects will not go into this level of complexity until later in the book but it is important to note that you are not always just cutting away a part from stock material.

Cutting foam and composites

Cutting foam is a very interesting way of using a CNC machine. You can shape a block of dense foam (available at most DIY stores) with a CNC machine and then lay-up fiberglass on it to create a very strong and light object. I have built parts of a full-sized airplane this way. The wings have a foam core that was then layered in fiberglass, but the shape was cut by machine (either hot wire or CNC). When cutting foam, you must have a high-speed spindle (10,000 RPM comes to mind) and very sharp bits to prevent shredding.

Foam also makes a *lot* of mess and it is always a good idea to mount a dust shoe/brush and hook the other end to a shop vacuum to suck up as much of the dust as possible. Wear a mask when cutting foam as you also don't want any of that dust in your lungs. One use case for foam is to cut inserts into packing cases and enclosures (for example, a camera case). Make sure your cutting lines overlap in your toolpath files so that you don't have any material that you have to cut manually afterwards.

Like foam, composites such as fiberglass also generate harmful dust. I use my CNC machine to cut fiberglass fins for my larger model rockets. The sheet material is about 3mm thick and the 3018 cuts complex geometry fins beautifully. Like foam, fiberglass does not like heat and will deform if it gets too hot (and your cut edges will look shredded). Fiberglass and carbon fiber use special end mills that look like deburring tools. Do this sparingly with your little 3018 because composites are very tough and can take their toll on your machine. If you only need one simple part, cut it with a band or scroll saw instead of using your CNC machine as that is overkill for something like that.

Secure your workpiece to your worktable very well when using the 3018. I generally drill holes in the stock material to screw it into the waste board, but because I am mostly cutting sheet material, I prefer to not let my cutting lines overlap. Instead, I leave little tabs that allow me to snip off the part from the stock material and then sand or file off the tiny bit of excess afterwards.

Cutting plastics and PVC

When cutting plastics such as styrene and acrylic you must consider the effects of having a bit or a laser penetrating the material and marring surfaces. Cutting plastics requires some aspects that don't apply to metals or wood. Many plastics including acrylic and plexiglass are brittle and will crack under stress. Therefore, clamping the stock material so that it cannot move will be important especially as loads are applied during cutting. Also important is the use of lower spindle speeds. High RPMs will heat up the surrounding material and cause the plastic to deform, which can add stress to it and ultimately result in cracks and breaks when least expected.

Cutting thicker stock material is much easier than thin. Thicker stock will hold its shape better and is less likely to get damaged by the bit. When I am cutting thin plastics (2 mm thick or less), I usually resort to the laser. By the same token, cutting softer plastics presents similar issues as thin stock material and is much harder to cut with a CNC bit. Again, where appropriate, I will revert to a laser for my 2D cuts. Because plastics are softer than wood or metal, you will need to have extremely sharp bits to cut through them. Dull bits will raise the temperature of the material around the cut line and give imprecise, jagged, and sometimes *remelted* edges.

Lasers present another dimension in safety consideration. PVC is a type of plastic that is very common and you probably have a lot of it as plumbing for your house. However, PVC, when heated, generates chlorine gas, which is poisonous. It has even been used as a weapon in chemical warfare (`https://en.wikipedia.org/wiki/Chemical_weapons_in_World_War_I#Gases_used`) so you can imagine that I firmly advise against cutting PVC with a laser unless your entire machine is in an enclosure with active ventilation to the outside.

For our purposes, the work we are doing with the machines we are using preclude us from cutting PVC with a laser. That is not to say CNC cutting is also hazard free. Like any plastic dust, PVC dust is hazardous and is best vacuumed up as it is generated. I frequently wear a mask with filtered vents to avoid inhaling any of that stuff.

Finally, let me just add a few notes on using lasers with acrylic. Acrylic can be cut with a laser but you have to consider the following:

- Transparent acrylic is a material that will just let light pass through where it will diffuse. You will then find that the cut is on the bottom surface (if the laser is not properly focused) or below the top surface of your material. For the most part, low-power lasers will cut acrylic but the laser has to go slow (4 mm/s) and multiple passes may be needed. Finally, air assist will likely be necessary. I would also add that acrylic smells terrible when it is being cut so do this in a well-ventilated area. Many acrylics are also semi-transparent, and others can be opaque. Cutting any material that allows light through will prove challenging with a laser without additional considerations (perhaps having some light-absorbing material stuck on top, such as the backing most acrylic sheets come with).

- Cast acrylic is formed by pouring molten material into a mold (think awards and things like that). This material frosts when hit with a laser and doesn't cut as cleanly. Use the spindle to cut this material.

- Extruded acrylic is typically mass produced and is generally cheaper and more common. Lasers will cut through it cleanly, but engraving will be clear versus frosted.

- When cutting clear material, I generally place some tape or something similar to prevent diffusion of the beam into the plastic. This also means I generally need higher-power lasers (e.g., 5-10 W) to cut through 3 mm or so. I have used a 10-W laser to cut through some of the clear stuff you buy at the DIY store, but that required really slow travel speeds because at high speeds the laser just went through and melted the material on the bottom surface instead. Also, multiple passes were needed.

- Thicker materials may require you to repeatedly adjust the focus of the laser if using a machine with no Z adjustment. Otherwise, you will need to adjust the position of the laser head by a small amount with each pass or series of passes.

Figure 3.1 – An example of a engraved piece of acrylic. This is very close
up to the actual engraving, which is why it appears rough.

Dust, debris, and gasses

When cutting, whether using CNC or a laser, byproducts are generated. With CNC, you generate dust and chips, while with a laser you generate gasses as material is vaporized. It is important to know the properties of each material when exposed to these cutting methods. This is why we will discuss upgrades to your machines such as the following:

- *Dust shoes*: These allow you to mount a brush to sweep debris and then vacuum it up through a hose mounted right next to the spindle.

- *Air assist*: For lasers, this blows air at the workpiece to keep any smoke from clouding the laser lens(es) as well as to push any harmful gasses away from the machine. You can have air assist for a spindle, but its purpose here is to cool the workpiece as it is being cut rather than clean up debris. You can also mist the material as you cut it to help keep the temperatures down. This is particularly the case for cutting plastics.

Cutting aluminum and other soft metals

Cutting through metals requires strong end-mill bits as well as the need to keep the material cool. If you have ever used a drill press on a block or thick piece of metal, you will likely have made the mistake of scorching the bit (where it gets black and starts to squeal as it cuts). This is when the bit gets very hot and binds against the metal it is cutting. This can be addressed with some cutting oil or similar coolant. There are tons of videos on YouTube showing industrial CNC machines cutting into blocks to make things such as engine blocks, or you might have seen lathes milling metal as well. In both cases, you will see the workpiece being drenched in a liquid that has two purposes. First is to keep the end mill and material cool and the second is to carry away the chips, dust, and other debris away from the work area so those don't interfere with the machining work.

For our desktop mills we have to rely on our dust shoes to collect anything the vacuum can pick up, but we also have to occasionally squirt a little machining oil as the bit is traversing the workpiece. The oil has a number of purposes: it keeps the bit cool by coating it, it keeps larger chips the vacuum didn't pick up congealed in the oil, and it also keeps the bit lubricated as it makes any holes it needs to make in the workpiece.

Our 3018 machines are not really suited for hard metals, and at best, we can machine small blocks of aluminum. If I needed to address harder metals, I would need a machine with more power and a larger spindle. However, our desktop machines should have no trouble doing some very light surface-level engraving on harder metals, although my biggest concern would be rigidity. I would not use the 3018 to actually cut steel as the machine is just not strong enough to handle metals like that and the number of passes required makes it uneconomical to use for small parts. If you choose to use it for harder metals the travel speeds will have to be very slow and you should keep the spindle and workpiece cooled with oil.

Cutting aluminum, however, is very workable with the 3018 and one task I have for mine is to fabricate parts out of aluminum to replace *BumbleBee's* yellow plastic pieces I have printed. I would likely have to revisit some of the holes post-fabrication to countersink them, but that's a small detail.

Aluminum comes in multiple alloys, each of which have specific properties. Another nice aspect of this metal is that the CNC machine will cut sheet metal (1-3 mm) very nicely. When machining a thin plate on the 3018, take care to secure the workpiece and consider that small, thin parts can deform while being cut. Generally, I like to leave any hole drilling for small thin metal parts as post-processing work rather than have the machine drill those holes for me. On the other hand, if I have a 10-mm block material that needs machining, then the 3018 will take care of that very easily.

Selecting the right end mill

A common theme so far has been that the end mill should be as sharp as possible. While they certainly wear differently depending on what and how they are made as well as what materials they have been used with and at what speeds, you should still consider your bits as consumables. Most kit-built 3018 machines will come with a handful of end mills (some as few as two and at least one kit I built had none). You should also consider different collets to accommodate the various end mills. A collet is the flange or sleeve that you insert into the spindle end that holds the end mill in place. On a common hand drill, this translates to the chuck, which you tighten by hand with a key. Other machines that use collets include Dremel tools. Here's a picture of some collets that came with my spindle; note how the hole diameters differ. This allows you to use different bits in your machine:

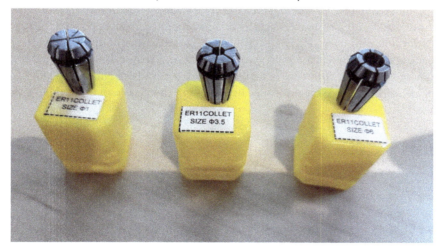

Figure 3.2 – Sample collets

Different end mills will have different diameters. This is why you need different collets. When you select an end mill for a task, ensure that it is suitable for the material as well. As mentioned already, consider whether you will be cutting out a contour (milling a shape out of a block of material) and whether you need an upcut, downcut, or compression bit. If you are milling aluminum you will want to select a bit with two or three flutes for chip removal since chips will pack in narrow flutes, and for wood a two-flute bit will do fine. End mills also have specific shapes for the task you want them to do and their length is important to address the depth of your cut.

Here are two frequently used end mills that I either purchased or came with the 3018. Note their different diameters. Some fine engraving bits are very thin and fragile and others have varying tips and flutes:

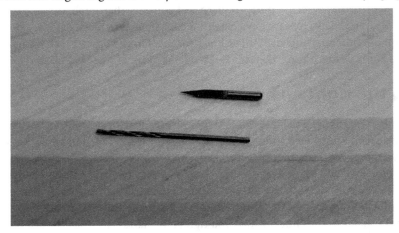

Figure 3.3 – Sample end mills: the top one is an engraving V-bit and the second is a two-flute flat nose

As you can see, different bits have differing purposes. For example, square-tipped end mills are very good for use with wood. Ballnose end mills are great for slotting and creating pockets, while taper end mills are good for holes, grooves, or angled edges. As discussed earlier in this section, chip removal with a dust shoe and vacuum and air assist or machine oil help ensure a successful cut every time. As you become more familiar with your machine, you will learn which end mills work best for the materials you are using, and we will cover specific bits in our projects so that you can see the effects of several of these.

Summary

We have discussed the handling of various materials in CNC in this chapter, and we have also touched on the various types of end mills. We will get deeper into a number of materials and end mills later in the book to give you some examples of specific tasks and materials. The number of permutations will boggle the mind so we can't cover every single one here, but I can detail a few and then give you the skillset you need to make your own decisions for your own projects.

As with everything else, all tools, including end mills, are not created equal. I cannot endorse any particular vendor in this book, but you only need to choose your end mill type and read the reviews from other machinists to make an informed decision.

In the next chapter, you will make inroads with your machine and start making some cuts, applying what you have learned so far before jumping into a number of projects that will build your skills as well as offer you the ability to upgrade your machine.

4
Making the First Cut

Now that our CNC machine is operational and provisioned, we can begin to test out its capabilities through some basic cuts. Depending on the material and the nature of our project, we can choose to cut or engrave – or do both. For example, we might want to make an engraved plaque but then have the CNC machine shape it out of our work material. We might also want to machine a component that has grooves, countersunk holes, or markings.

This chapter will show you how to do the following:

- Secure your workpiece especially if your ultimate objective is to end up with a finished product cut out of the stock material.

- Select a test pattern to see the various effects. One of the options here is to use a laser to replicate shading effects as might be visible in a photograph. Think of a black and white image and the various shades of gray. You can replicate this using a laser on just about any material.

- Manipulate configuration settings to achieve specific goals. For example, travel speed is impacted by the material and whether the intent is to cut all the way through the material or not.

Configuration settings are also impacted by the end mill you are using. If you are dragging a spinning bit over hard material, you want to ensure accuracy. So, how fast your toolhead moves at a specific depth is important so that you don't expose your end mill to bending forces that not only make your cuts inaccurate but could also result in the end mill breaking. Another consideration is the diameter of your end mill. Fine lines need thinner bits, but if it is not practical to swap out end mills mid-run, you will likely need to contend with making multiple passes to engrave wider lines. The same would be true for cutting through thicker materials with small end mills. Your machine may have to gradually carve its way through the material. Think about this as well: cutting generates heat (have you ever seen how hot a drill bit can get when going through plate metal?), and if that builds up on your material you could char it (if it's a composite such as wood or MDF) or deform it (if the material is acrylic or plastic).

Securing the workpiece

My CNC machines all have tables made of aluminum extrusions, except for one that uses a piece of plate metal. Both would get ruined if I made a mistake and cut through my workpiece and into the table. Eventually the structure would weaken, and I would have to disassemble the entire Y-axis assembly and replace the table – not something I would relish having to do multiple times. Consequently, as has been mentioned before, I like to put a piece of waste board between my workpiece and the table. The waste board is typically made of MDF or plywood and is sacrificial. However, with a thick piece material on top of my table and under the material I am trying to cut or carve, I have to find a way to bolt the material to the waste board and the waste board to the table.

You can purchase waste boards for your 3018 on Amazon or AliExpress easily enough. These often come with countersunk threaded inserts as well, so that you can secure your board to the table without having a screw sticking up above its surface. Even if there was nothing provided, you could simply clamp the waste board to your table using binder clips. You would just have to make sure your end mill doesn't strike the clips.

The threaded inserts now provide a way to screw in a bolt and clamp assembly to hold your workpiece over the waste board. Most CNC machines come with a handful of clamps of various kinds. Some are nothing more than a large bolt with a piece of steel through which the bolt can go through. Tightening the bolt to clamp the steel onto the workpiece will hold it in place. Of course, you would again have to be sure that your toolhead does not collide with the clamp because that will most assuredly wreck the end-mill bit.

Sometimes, all you need is just a good old-fashioned clamp. Here is an example of one of my smaller lasers cutting into a piece of acrylic. Notice, at the bottom right, the clamp holding the workpiece in place so that it doesn't move while the laser cuts into it.

Figure 4.1 – Small laser cutting into some acrylic clamped to the
table with an ordinary clamp (bottom right)

However, a clamp is not always the sole solution. If you are cutting a part out a piece of stock material, once the part separates from the base workpiece it can *float* loosely, get hit, and then be damaged by the end mill. There are several viable options to keep your workpiece safely in place:

- Sequence the cutting process to allow you to use any holes that may have been cut to secure the part with a screw into the waste board – note that the screw head will likely have to be countersunk. You could even incorporate this into your object's design. For example, let's say your design calls for two or three holes along the length of the finished product. Have your machine cut those holes out first and then put a screw into each hole and into the waste board underneath. Now cut the rest. Your part will release from the stock material but be firmly secured to the waste board and protected from inadvertent movement.

- Define a toolpath that avoids cutting through all the way around a part, leaving some tabs to keep the finished item attached to the source material, and allowing you to separate the part from the stock with hand tools. If your material is thick enough, you could have your end mill cut down to the last millimeter or so and leave the tabs thin enough to snip or saw off with small hand tools. Years ago, you could buy balsa model airplanes where the parts would be die cut into a sheet of balsa. The die would leave tabs so that parts remain attached to the sheet, but still allow you to finally separate the part using a small sharp knife to cut the tabs.

- Another option is to use vacuum holding. This is where suction holds your workpiece in place against the waste board. The waste board surface has to be very flat and free of debris. Of course, with our little 3018s, using a vacuum to hold our workpiece will likely not be very economical.

- I have heard of some people who also use double-sided tape to hold down the stock material. Another alternative is to use spray-on contact adhesive. This may not work so well with heavier materials and small parts that may be fragile and difficult to remove.

- Finally, there are other products such as are heat-activated adhesive substrates meant to easily peel away. I have not had the opportunity to use them as the basic mechanical methods of securing my materials have been more than sufficient.

In the following figure, you can see an example waste board. While relatively inexpensive, you can also make your own with some MDF and a drill press. Then a few threaded inserts can be screwed in to allow for clamping your workpiece directly to the waste board.

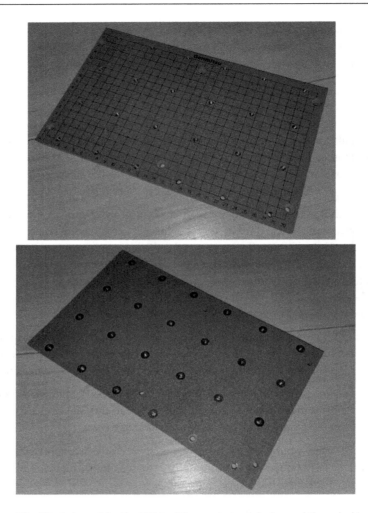

Figure 4.2 – Waste board for the 3018 with countersunk holes and threaded inserts

I bought this board online, and I have another machine using something like this. However, I also have purchased some of those inserts that you see in the preceding photo in case I have to cut my own board. You can easily use a store-bought unit as a template and make as many of these as you want.

Selecting test patterns

Whenever I set up a new CNC machine, I attempt to cut a series of basic shapes first and then work my way up to more complex tasks. I prefer to start with some simple carving into MDF or plywood. I try to avoid cutting through the material because I want to test the machine against the material and get familiar with how well it handles certain tasks. Not all machines (or end mills) are created equal, and this is the time where I look for obvious limitations.

Choose your test material and bolt it down to your waste board. I typically start with something that is 20-25 mm (up to 1") thick or more and set my design to a cut depth of 1 mm or so, gradually increasing the depth of my cuts to see how well they come out. I am effectively carving or engraving here, and for our purposes that's good enough. You might want to prepare several small test boards so that you can use them to calibrate your machine. By cutting only to an initial depth of 1 mm, you can flip the workpiece over and dig into the other side, still leaving core material between surfaces.

I like to run my tests against a series of basic shapes, such as triangles, circles, or rectangles. You can draw these in your favorite design application, or you can use patterns that are available in the public domain. *Inventables* is a firm that caters to hobby CNC machinists and publishes a great test pattern you can try at `https://www.inventables.com/projects/calibration-test-pattern`. Besides putting your machine through its paces, the purpose of the test pattern is to also validate the operation of your stepper motors. If your motors are not getting enough power, the firmware is not set to support the required steps, or your motors are uncalibrated, your motor(s) may miss steps and the CNC machine will produce unpredictable results. The test pattern will also test the alignment of your toolhead and more specifically how perfectly perpendicular your end mill is relative to the worktable and workpiece (it has to be absolutely perpendicular).

Load the test pattern into UGS or your desired sender and let your 3018 cut through it. The pattern from Inventables doesn't just exercise the cutting of shapes, but also works through cutting at various depths for these shapes. The test pattern is a great way to see how well your machine handles different shapes as you cut different materials and use different end mills. There will be many permutations and settings that work for you and your particular setup, which may or may not match my own machine settings.

Configuration settings

The settings for any given project or cut will revolve around several parameters. However, in this section, let's review some basic settings I use to smoke-test every workpiece.

End-mill diameter

You must always determine how much room your toolhead has to move between lines. If you have two cut lines that are very close to each other and the diameter of the end mill will span that gap, a design revision may be needed, or a revised selection of end mill or material will be necessary. Generally speaking, this will be apparent if you look at your toolpath and determine that two cut lines come too close together. Of course, the depth of your cut plays a role here as well. If we have a non-uniform end mill (where the cut width will change with the depth) then the toolpath is something to consider. For most of what I do, I stick with a few common bits for my 3018 machine and so I do not change the bit radius/diameter once I have set it. If I am using UGS, I like to visualize the cutting process, so I get an idea of what to expect when the machine gets to work. I don't necessarily need to see the entire process, but typically I look at how close the lines will be and whether there might be any issues moving around complex shapes. Here's how you launch the visualizer in UGS:

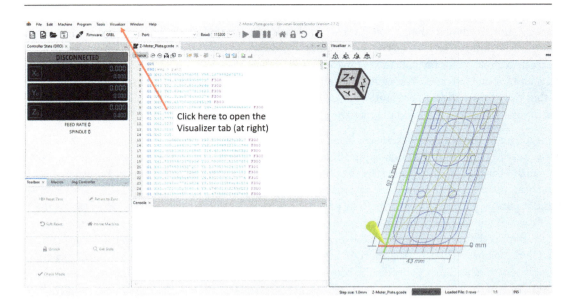

Figure 4.3 – UGS Visualizer button

Use the **Visualizer** button in UGS to see the anticipated toolpath before you do any cutting and adjust either the design or your G-code as you see fit.

RPM or spindle control and feed rate

We have already touched on this to some degree. The RPM setting for your spindle must be sufficient to cut the material but also to not subject the bit to enough heat to scorch the bit or the workpiece. Feed rate is another term for travel speed. You will see that in UGS as well. On some 3018s, the RPM can't be controlled by software, and for those, a PWM RPM controller can be mounted between your spindle and the controller board. However, nowadays, you should be able to control the spindle just fine from within GRBL and UGS. For example, the S12000 M3 command will set the spindle speed to 12,000 RPM. I highly recommend you test your cut on a piece of scrap and use that to play with settings for the material. Every workpiece is different and once you know how the properties of your workpiece material work with your machine, you can develop profiles for each material you put through your 3018.

Cut depth and multiple passes

I have found it sometimes impractical to cut through material (such as steel) in one pass with my machine. The bit may get too hot and require additional cooling through air assist. So, rather than try to get the end mill to plunge into my piece of metal, I have it scrape off layers by going over the toolpath multiple times. You will find this is common when cutting straight through with a laser, but with CNC, it's a somewhat different animal because you may also be shaping your workpiece (you may have features at different heights, or countersunk holes, for example).

The cut depth or the depth per pass can be set in G-code. When you generate the G-code from your design, you should be able to define the depth per pass and so the number of passes. This tells the 3018 to move across your workpiece as many times as needed at different Z heights. There are calculations you can make around what the depth per pass should be based on your end mill, and those depend on the bit, the material, and the cutting-edge geometry, expressed in a co-efficient. The formula is typically the radius of the bit multiplied by this co-efficient. That information may not always be available, so there is another, simpler rule you can use: your cut should not go deeper than one-half the radius of your end mill. Therefore, if you have a 0.125" radius bit, the depth per pass should not be more than half the radius, or 0.0625". With some experimentation you might find variations that work for specific materials as your machine can handle them. If you experiment, be sure to increase depths in small increments so that you do not risk destroying your end-mill bit.

Different settings for different operations

We have already touched on some of the considerations around settings related to not just what you are cutting, but also how. For a laser, multiple passes where the beam is likely to diffuse inside acrylic may mean wider cut lines and multiple passes (acrylic tends to re-melt if the cut line is thin, requiring higher power beams and multiple passes). When using an end mill most of your settings will be in the G-code generator program. However, there are some basics I have found that serve me well.

If I am engraving wood, I keep the depth to 1-2 mm at most. If I need shading and changes in texture, I will switch to a laser. If I am engraving metal with an end mill I never get deeper than 1 mm and prefer to stay in the realm of 0.2-0.5 mm.

I have broken enough bits trying to find a baseline feed rate for every material I typically work with. This feed rate is calculated based on the bit, the RPM, the material type, and so on. You may find the feed rate also defined as the **surface feet per minute** (SFM) or **surface meters per minute** (SMM) – this is, again, the rate at which your toolhead is traveling and cutting your workpiece. Chip load is the amount of material your end mill removes with each revolution. These two numbers are often defined for you by the end mill manufacturer. The feed rate is calculated like this:

Feed Rate = Spindle RPM x Number of Flutes (or teeth) in your bit x Chip Load

As you can see, the chip load is the biggest governing factor, and it is material and bit-diameter dependent. Most machinists memorize feed rates developed through experience. I like to set my spindle RPM to between 12,000 and 24,000 RPM and use these chip load ranges as a baseline for 3 mm diameter bits (I have derived these from my reading over the years):

- **Hardwood**: 0.08 to 0.13
- **Plywood**: 0.1 to 0.15
- **MDF**: 0.1 to 0.18
- **Aluminum**: 0.05 to 0.1

Irrespective of equations, nothing replaces experience in determining a suitable feed rate for a given material and diameter bit. I drew these values from a variety of sources, but you might find that different types of woods or even alloys demand different chip load RPM settings or feed rates. Sometimes you might find end-mill vendors providing details on acceptable feed rates for their product. Only use a formula when you have no other avenue because you do not know the material well enough or are unsure of the end mill itself. Here is an example calculation for a two-flute flat-head end mill running at 12,000 RPM on plywood:

Feed Rate = 12000 x 2 x 0.1 = 2400 mm / minute

There are several feed rate calculators on the internet that do not make use of chip load numbers, and even with specific material chip loads, these can vary greatly. Once again, experience will inform you best.

I have no doubt you will break some bits trying to find a sweet spot for each bit size and workpiece material specific to your machine. Just keep in mind that thinner bits mean slower feed rates and RPMs. Also remember that slower feed rates do not mean your cut will fail, but that they will take longer to complete. Most of the time you will be able to use your G-code sender program to adjust the RPM and feed rate as your cuts progress. Listen to the sound your cutter makes, and you will know whether a failure is imminent or you can tweak your settings a little to get more performance.

Summary

The focus of this chapter was on some basic settings for your CNC machine and how to optimize them. We have also learned how to confirm our machine is operating to expectations by running some test patterns. As we progress and your experience with your machine grows, you will develop your own test workflow as well as optimum settings for your projects, materials, and end mills. There are many variations here because not all of our 3018 machines are the same (and if you are using something other than a 3018, you definitely have your own tweaking to do).

The next chapter is the opener for a series of special projects. We will have the opportunity to try our hand at setting cut depths and/or depths per pass and look at how we might make the same *thing* out of different materials. This will allow us to investigate the settings we have discussed here and see what the effects are when we cut soft wood, hardwood, foam, acrylic, and other materials.

5
Full CNC Workflow with Different Materials

We are ready to move forward with milling various materials beyond simple testing. It is important at this point to have our machines calibrated and able to accept G-Code on an ongoing basis. As mentioned several times before, the type of material and the nature of the shape dictate the settings and how well our cuts come out. For every project there will be common and specific settings, tweaks, and other attributes that may be unique to the material, the machine, or the project itself.

Here's where we will focus our attention in this chapter:

- Converting a non-CAD drawing into something your 3018 can process – a full workflow
- Working with soft wood that requires a delicate touch since the resulting parts may be fragile
- Working with soft metal, such as aluminum
- Working with hardwood (our example will involve attempting to engrave/carve a piece of 2x4 or some similar material)
- Understanding how to work with dense foam
- A little more about setting depths and engraving

I recognize that you won't always be cutting, but might be engraving, so it bears spending some time talking about it in this chapter. Another notable difference from other discussions earlier in this book is that we've left acrylics out of this conversation. I prefer to cut acrylic with a high-power laser, but engraving it is easy with a low-power unit. When I do cut acrylic with the 3018, I like to be careful to ensure there is no chance of splintering or remelting by going across the material multiple times, slowly, and using wide lines so that there is a gap between then. Needless to say, when engraving, that is less of an issue.

Technical requirements

In this section, we are going to use several tools. You are not constrained to these tools specifically, but the idea is to apply what is out there to your needs. I generally like to have the following:

- Something that will convert an image to a vector graphic. This helps me make use of hand-drawn or scanned images that I would otherwise have to re-work. You would need to gauge whether the amount of re-work you must do is worth the effort or whether it is better to just redraw an object. I particularly use this kind of tool whenever dimensions are in short supply, or the scale is not clear or workable. For my BD-1 project, the drawing arrived scanned but also with some smudging on the lines.

- If you work with CNC, you will eventually have to learn how to use CAD to either create or change existing drawings. There are several free applications out there. I primarily use TinkerCAD, but there are others (free and not free), such as FreeCAD, Fusion 360, OnShape, and many more. Pick one that will export files in a workable format for you. Some will even generate G-Code natively, which would eliminate the next tool I have in my list.

- I also like to have another file converter that will take the vector graphic and convert it to G-Code. As mentioned, if you have a CAD tool that can import vector graphics and generate G-Code, you don't need this. Some tools are better than others, so do experiment.

One great aspect of using tools like these is that you can scale anything as you see fit. And if you need to test the fit and scale of parts, you can make any changes as required before you cut anything.

Getting to G-Code from a drawing

One of my other hobbies is model airplane building and while I have mostly focused on static scale models, I also enjoy building RC airplanes, but I like to do so from plans, and most of the airplanes I like are not readily available in kit form and the plans themselves are not always clear or easy to read. A substantial effort has to be made to transfer the drawing to electronic form, scale it correctly, and then reliably pass it on to the automation to cut the part. One common class of parts are fuselage bulkheads. For small-scale models, balsa is called for, while larger models would do fine with plywood.

Here's the bulkhead from the scanned BD-1 blueprints. Note that there are no measurements, so you would have to use a ruler to take your own – assuming your drawing will print properly at full scale.

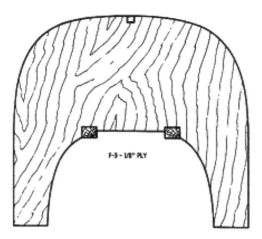

Figure 5.1 – BD-1 scale model bulkhead

The first (full-sized) airplane I ever owned was a Grumman American AA-1A. The AA-1A is a development of Jim Bede's BD-1 which is, for the most part, the same airplane as my AA-1A in appearance (obviously there were changes in structure, engine, and other attributes as the design was developed). So of course, this model is very nostalgic for me. You can see a picture of the airplane (I affectionately used to call it the Flying Pumpkin because of the paint scheme I painted on it) at https://www.jetphotos.com/registration/N9232L. I sold the airplane many years ago and recall many adventures in it, including flying it in the UK as well as up and down New England, Maryland, and Virginia.

If you look at the bulkhead in *Figure 5.1*, you will see that three notches are needed. One is at the top of the bulkhead and two more are on the interior of the curve at bottom to allow the bulkhead to sit on a pair of stringers.

So how do we convert what we see in *Figure 5.1* into something that is machine readable? Let us find out by going through the following steps:

1. First, cut and paste the image into something like Microsoft Paint so you can save it as a .png or .jpg file.

2. Convert the image into a **Scalable Vector Graphic** (**SVG**) file using an online converter. Any suitable converter will do. You can use Adobe if you have a license, or Convertio (https://convertio.co/jpg-svg/). You will now have a transferable, albeit suboptimal, image.

3. Import the image into a CAD system such as TinkerCAD. This is very straightforward in TinkerCAD. You create a new design, click the **Import** button, and then point to the SVG file to load it in.

4. Now resize the image according to the scale on the plans. Also ensure that the CAD drawing is the same thickness as indicated on the original plans (all properly-done plans show the scale in the plan legend). Sometimes this is represented as a checkered bar indicating how many inches (or meters, or feet) the bar represents, others represent scale as a ratio, for example 1" = 24" or 1:24. Here's the scale on my plans:

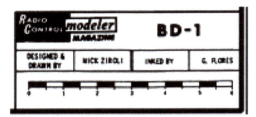

Figure 5.2 – Scale and legend from my BD-1 Plans

In the absence of any information, pick a dimension on an overall drawing, such as for the whole airplane, and measure that, and then compare that against the real thing to get a scale ratio. For example, for my airplane, I know the actual wingspan and so can measure the length of the wings on the drawing and determine the scale of the drawing. The following figure shows you the resulting CAD drawing:

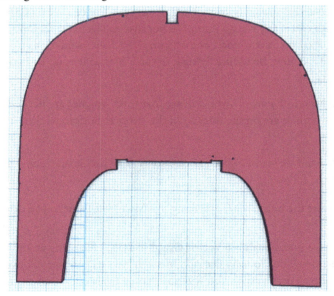

Figure 5.3 – BD-1 bulkhead converted to a CAD drawing

5. You can now generate G-Code out of *Figure 5.2*. You can use an online utility such as *FileStar* to convert the SVG file into G-Code and then load that into our trusty G-Code sender, UGS. The following should be the result:

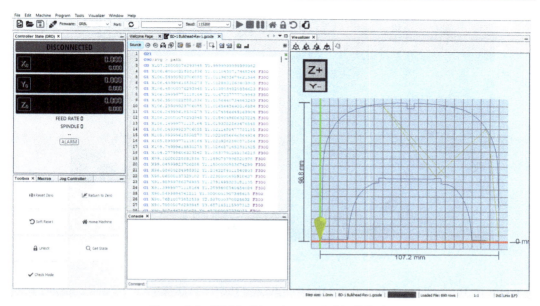

Figure 5.4 – BD-1 bulkhead converted to G-Code

Despite all these gyrations, you can see we have a path from a drawing in PDF form (that was itself originally hand-drawn) to a G-Code file ready to send to the 3018 machine. We can even scale our image as much as we like before we generate the G-Code, which allows us to prototype all we want. With all this, we can now move forward and cut the bulkhead out of light plywood.

Cutting softwoods such as balsa and light plywood

For our purposes, I am using 3 mm plywood to cut out my bulkhead because most of the balsa wood in my shop is smaller planks for park-flyer-type (i.e., smaller scale) airplanes. I use the same plywood for model rocket fins, and while light and easy to mill and machine, it is very durable.

Both balsa and plywood are going to chip on the "bottom" surface (the surface facing the wasteboard) if we are not careful with what endmill we use. To prevent the "tearing out" that would happen with plywood, I recommend a flat-head downcut endmill. I am using a single-flute flat-nosed endmill with a 1/8" shank. These are commonly available from various vendors, and you should take into account that you will likely break a few as a matter of course. I like to keep my feed rate slow, at or below 1000 mm/s for thicker plywood and 2000 mm/s or less for thicker material. For the spindle, keep it below 24,000 RPM.

Here is a picture of my machine carving out the bulkhead. The plywood came from my local DIY store; it is mostly scrap material that has provided panels for various projects as well as a painting platform.

Figure 5.5 – My 3018 carving out the BD-1 bulkhead

When cutting through balsa, slow down the spindle and the feed rate even more and consider the effect of travel along the grain of the wood. There, the feed rate will meet less resistance and so can be faster. For light-duty plywood and balsa, you can make multiple passes so that you don't tear up the material. As you will have read already, you can leave in tabs on the outermost perimeter so that your part doesn't detach from the stock workpiece or secure the part to the wasteboard with a screw. For my purposes, I made two or three passes, screwing the part down only after the first or second pass (which allows me to see where the spindle is not going to be impacted) because I did not want to have to cut through 3mm tabs. If I was using balsa, I would use tabs and cut through with a sharp knife.

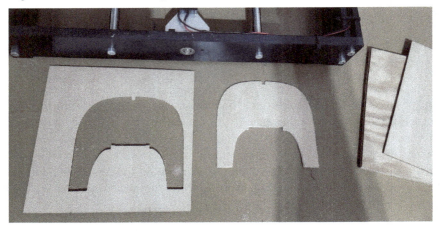

Figure 5.6 – BD-1 bulkhead cut out of plywood. Note the original sheet of plywood showing the cutout

Learning to use the 3018 to cut plywood is a good way to learn how to cut most wood sheet. The next section touches on cutting hardwood, and in my case, it concerns recycling materials I already had.

Cutting hardwood

Late this summer, my wife decided that the wooden table we had on our deck had outlived its purpose. Even though it was meant to be an outdoor table, it made it all of 2 years before the wood started to rot in some areas. Consequently, it had to go. So, I got to take it apart and since I did not really want it to go to the dump, I cut it up into pieces small enough to put into the trash or recycling. This is when I discovered the table's legs were made of this exquisite and still good hardwood. Naturally, they encountered my miter saw and now, slices of the table's legs are workpieces for my CNC machine and laser cutter.

The denser the material you cut, the more your machine needs to be rigid and your axes precise. As you gain more experience with various materials, you will find the feed rate for the denser materials, such as hardwood, will have to be slower and require multiple passes. As before, we are going to use a downcut bit because we want the bit to push down on the wood to minimize splintering. For hardwoods, the spindle RPM can be on the lower side (between 10,000 to 12,000 RPM) and your feed rate can be between 35in/min to 40in/min assuming a 0.125" (3mm) thick piece of material. However, your mileage will vary based on the specific endmill (these preceding values are for a two-flute bit).

Cutting and engraving soft metals

Metals present a similar challenge to cutting hardwoods. Your endmill has to go through stronger material with the added problem that the metal-on-metal contact can generate lots of heat (you will generate heat with hardwood too, but more so with metal). When you cut metal, you can use air assist to keep your bit cool or keep spreading lubricant on your workpiece. Either way, if you thought cutting wood created a mess, cutting metal creates an even bigger mess. This is why I use a 3018 with a dust shoe hooked up to my shop vacuum whenever I am cutting metal. The 3018 is not really suited to cut large pieces of metal, but you can cut small parts so long as you go slow and make enough passes. For my purposes here, I am cutting a part to replace a plastic part on BumbleBee. The larger BumbleBee is meant to machine larger parts, but because some of its critical parts are made of plastic, it can't really handle metal. However, I can have the 3018 make some key parts that allow me to upgrade it.

Here's BumbleBee – all those yellow plastic parts would be better if they were made from 3 mm aluminum:

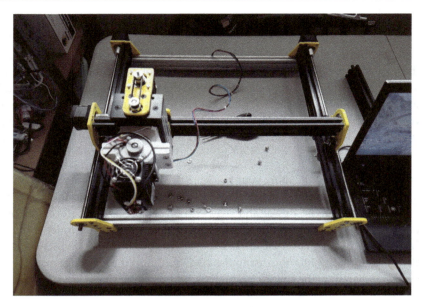

Figure 5.7 – BumbleBee in all its plastic-part wonder

One thing to consider is whether to pre- or post-drill holes on your workpiece. Doing this helps the machine not have to drill through the material and focus instead on cuts for countersinking or less straightforward machining associated with holes (e.g. creating nut-traps). There are several ways to do this. Here are two of them:

- Once you have set the origin, have the CNC machine make a single pass through your stock workpiece. The machine should define the location of any holes. Mark where your workpiece is on the CNC machine so you can put it back in the same place afterwards. Now remove the workpiece, drill your holes in a drill press, then put the workpiece back and run through your cut as many times as you need to. You can also use one or more of your predrilled holes to hold the piece to the table so that it doesn't move after you are done cutting it out of the stock material.

- Print the cut pattern on paper and glue it to the surface of the aluminum. Then predrill all your holes and place the workpiece on the CNC machine. Make sure you set the origin appropriately in G-Code so that your cuts will go where they are supposed to go. If you set the origin somewhere just outside the border of the outermost cuts, you can set that point on the workpiece every time and get the cuts to be exactly right every time.

Since you can set the origin in G-Code, you can use that as an identification point to line up your stock material on the CNC machine to address unique operations that the machine but not be well suited to do. If you have a 3018 with both a spindle and laser toolhead, you can use this origin to zero your machine for laser engraving as well. You can learn more about the benefits to setting your origin with this video: `https://youtu.be/A1zlL3q23HI`.

The part I want to cut is available as an STL from Thingiverse. I imported the STL into TinkerCAD and then exported it as is as an `.svg` file and from there, I converted it into G-Code just as we did earlier in this chapter. Here's the original part:

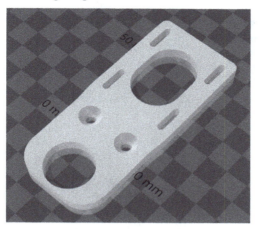

Figure 5.8 – BumbleBee's original part rendered for 3D printing

From there, UGS did the rest of the work. Here's the part as G-Code:

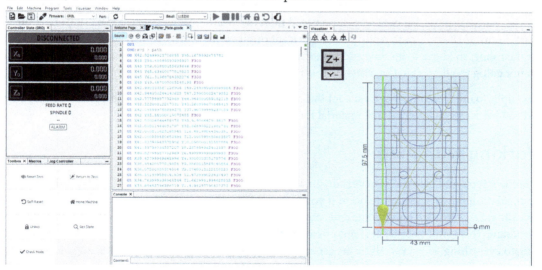

Figure 5.9 – BumbleBee's part rendered as G-Code

Because I have oblong holes, I didn't bother with pre-drilling. Here is a picture of the 3018 cutting the part out of a stock piece of 3 mm aluminum plate I have lying around. You can probably figure out where this part goes when you look at the BumbleBee photo shown previously.

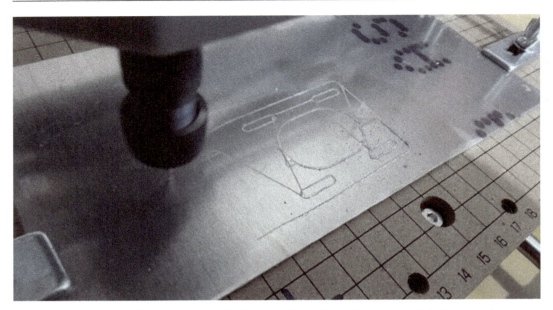

Figure 5.10 – Cutting a part for BumbleBee out of aluminum. You can see I failed to raise the
Z axis here while cutting – something to watch out for when converting your files

For this part, I kept the bit cool with slower feed rates (even slower than hardwood). I don't have air assist on my 3018 and I can't really pour a lubricant or oil on the workpiece because that would foul the dust shoe, so I have to go slow and make lots of passes. Obviously, if I only want to engrave and I am not going too deep into the material, this is less of a problem.

Next up is foam, which we can treat a lot like soft wood. Foam is fun to work with because it is light and can be used as a foundational material for very lightweight structures.

Working with foam

There is a type of dense foam material you can buy at your local DIY store that is typically used for insulation but is an absolute boon if you are an RC airplane or model enthusiast. The foam is typically called extruded polystyrene and comes either in pink or blue sheets. I am not a fan of milling styrofoam. This is not because it is not possible, it's just that it is much messier because bits of foam cling to everything. However, all foams generate dust that is very bad for your health. So, it is doubly important to have a vacuum dust collector on your machine and to operate the machine wearing a respirator. Make sure you clean up thoroughly afterwards. You might even consider keeping your machine inside an enclosure while it is cutting foam to maintain control over all the dust. Here's what dense foam sheet looks like:

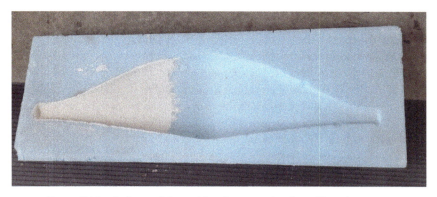

Figure 5.11 – A sheet of dense blue foam used to form fiberglass parts
for my airplane. Note the machined pattern in the foam

There is an entire segment of CNC that cuts foam parts using hot wire driven by a machine not unlike a CNC machine. Frequently though, smaller projects are cut manually with a handheld hot wire cutter and a cardboard or plywood template. Dense foam is really fun to work with because you can create complex geometries and the end product is very light. You can then lay the structure up with fiberglass and create a strong but very light object – ideal for rocket and RC airplane parts. Even the full-size airplane I am building uses this technique. The wings are CNC-cut out of a special dense foam that is then layered up with fiberglass.

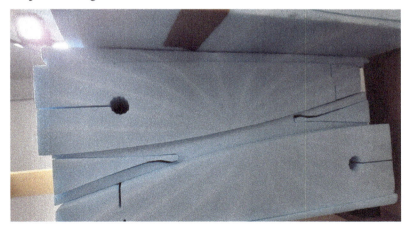

Figure 5.12 – Wing parts for my airplane cut out of dense blue foam. Notice the
originating block of foam is still there. These were cut with a hot wire foam cutter

When you are working with fiberglass, you will have to use a specific bit for your mill. Common wisdom is to use a bull nose, end burr/radius, or ball endmill. Of course, you will need to take the width of the bits into consideration for your toolpaths – for instance, if you have to open up a 2 mm hole, you will not be able to do it with a 3 mm ball endmill.

Let's start by examining some wing parts. The BD-1 wing is linear from root to tip. The airfoil profile is consistent all along the span of the wing. The original BD-1 (and its successor, the AA-1) had the wing spar made from an aluminum tube that also acted as the airplane's fuel tank. Since this is an RC airplane, I will still need to accommodate the spar, but happily no fuel needs to be stored in there. Just as with the bulkhead we start with the original drawing and in this case, it is a wing rib. We can cut as many of these as we want, slide them onto the spar, and then cover them with a light fiberglass to get a strong, solid wing. The scanned drawing leaves much to be desired, however, and ultimately, we would spend too much time trying to clean up the drawing.

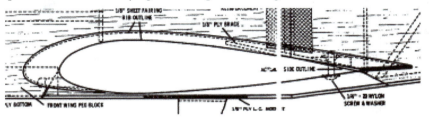

Figure 5.13 – BD-1 wing profile excerpted from my scanned drawings

The answer is to go back to TinkerCAD and hand-draw the airfoil, scale it accordingly, and then turn that into G-Code. The BD-1 used a NACA 64-415 airfoil, which I could replicate, but I went with a far simpler and more common Clark-Y. I found a good representative airfoil on Thingiverse (`https://www.thingiverse.com/thing:4192549`) with suitable holes and slots for spars that I could convert into a vector file, and from there into G-Code. The idea is to cut this profile out of thicker pieces of foam and just stack them on top of each other, and then run a carbon fiber rod through some of the holes or a beam made from plywood as the load-bearing wing spar. As before, I put the downloaded STL into TinkerCAD:

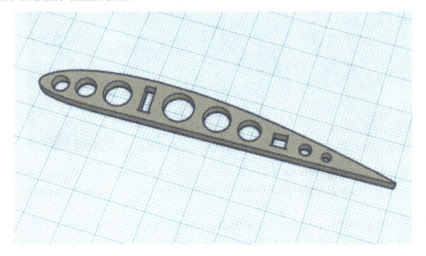

Figure 5.14 – Clark-Y airfoil

From there, it is a simple process to convert it into G-Code just as before: we convert the file with FileStar and end up with the G-Code we need to move forward:

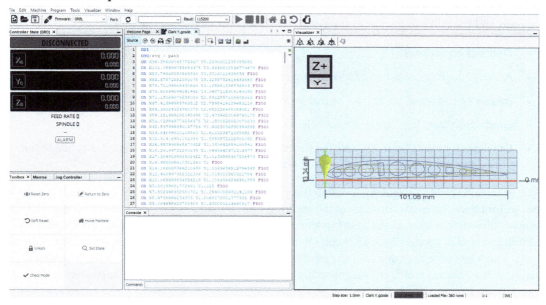

Figure 5.15 – Clark-Y airfoil ready to be sent to the CNC machine

Please note that the dimensions of this airfoil do not conform to the appropriately sized wing of a BD-1 whose bulkhead we cut earlier in the chapter. These steps are intended to simply illustrate the workflow, rather than giving you specific dimensions to build a specific object (in this case, a BD-1). However, if I wanted to scale my wing appropriately to the bulkhead, it is a simple process that can be done right in TinkerCAD.

The same can be done with large tranches of the fuselage of the airplane. If, for example, I wanted to model the completed fuselage, I could mill it out of foam, apply a suitable covering, and end up with a lightweight model that I can build and rebuild as many times as I want.

If you are interested in more model airplanes made out of foam and other materials, I would encourage you to have a look at the FliteTest website at https://www.flitetest.com/. They build and fly some amazing things.

Next, we will address conditions where you do not want to cut anything at all, but rather engrave an object with a pattern, either in 2D or 3D.

Engraving your workpiece and setting depths

Let's now tackle engraving your workpiece. Maybe you have cut your part on the 3018 and it looks awesome, and you'd like to mark it in some way. Here are some engraving use cases:

- Decorative purposes
- Adding a logo or some sort of identifying mark tying the part to you
- Inventory and serial numbers – so that you can tag a certain part
- Graduation markings (for example, ruler notches for measurement)
- Labeling (for example, to indicate the function of a control that might later be attached to the finished product, such as an indicator of where the on/off detents are)

If you are working with opaque materials, letting paint (or for wood, stain) seep into the engraved areas will enhance the visual attributes of the marking. On/off marks can be marked in red on a white surface, while ruler notches might be in black on a metallic piece. Engraving need not be done with an endmill, and in fact, I generally don't engrave with a bit if I can avoid it. This is because I generally use a laser, which gives me the same effect, is much faster, and does not "dig" into the material as much. Plus, I can have the laser etch out an engraving color (e.g., having it etch out the metal underneath an anodized surface).

You may still have reason to engrave with an endmill bit (maybe you don't have a laser or don't want one). Just as before, you can set the origin for your G-Code, but remember that unless you are swapping the spindle with your laser toolhead (i.e. the laser is mounted with the spindle), you will need to compensate for the offset of where the laser toolhead actually is relative to the spindle's centerline. Experiment to find what those settings are and simply run each process separately. Obviously, depending on where your laser is mounted, you will want to move the origin more to the left or right than what you had for the spindle. If you simply slide out the spindle and mount the laser in its place inside the same mounting, then the origins will be the same for both toolheads.

However, let's say you don't have a laser and need to engrave something on the surface. If that engraving can be done with the same endmill, simply include it in your G-Code but omit it after one or two passes if you must make multiple passes through your material. Essentially, you could have two files with different versions, one with the engraving and one without. Or, you could separate your work into two files, one that contains just the engraving and one that contains everything else. In any case, there is no reason for you to go too deep into the material and compromise it. Typically, I like to stick to whatever a single pass's depth entails. You only need to "scratch" the surface enough here to do what you need to do.

I would point out that a laser engraving on transparent or semi-transparent acrylics or materials is somewhat different because you are affecting the material below the surface. We will look into lasers and other CNC applications that don't involve removing material later in the book.

Summary

We've covered a lot of ground in this chapter. We invented a workflow to go from a scanned or hand-rendered drawing to a physical object. Congratulations! You now know how to turn a picture, or even just an idea, into something you can hold in your hand and actually do something with.

We also did some hands-on investigation into the effects of cutting different materials and saw what we needed to do to successfully complete a project, including doing some prototyping. At this point, you should be fairly comfortable using your machine to finish your projects. This is important because in the coming chapters, we will be exploring enhancements to our 3018s as well as investigating how we might CNC on a 4th axis.

6

Upgrading Your CNC Machine

With our machine now working well, it's time to consider what improvements we can implement to add to its functionality and make it work better. With some common components, we can improve our machines' performance and extend their capabilities. In addition, there are several add-ons that can be easily acquired or fabricated.

In this chapter, we will cover the following upgrades:

- Adding end-stops so that our machine can home itself and not require us to set the origin.
- Adding a Z probe so that we can define the starting point on any surface.
- Giving your machine an emergency STOP switch so you can prevent problems from getting worse without having to rely on the computer and risk lag when telling your 3018 to stop. This is also important if you intend to operate the machine untethered from a computer.
- Introducing the ability to carve and cut on a rotary axis.

We will also upgrade your 3018 to be able to operate as a plotter and as a drag knife.

Installing end-stops

If your machine has no way to home unless you set a specific origin, then you have no guarantee that your toolhead will always start at the exact same position as the origin every single time. Consider this scenario: you move your toolhead to a specific location on your surface, set that as the origin (in the software), and then start a cutting job. In the process, your machine gets jammed while moving. You then stop the machine and try again, starting by telling the machine to home. However, the machine *thinks* it is where it should be, but in fact isn't. Consequently, when it tries to backtrack and return to what it thinks is the origin, it will instead reposition the toolhead somewhere else altogether. This problem is also encountered when you need to cut the same object multiple times. Having to keep setting the origin manually can get tedious and will make it annoying to re-attempt a job because you have to mark the origin you select somehow and bring the machine back to it every time. Why bother, when you have automation take care of that for you?

You can install as many as six end-stops on your 3018, but unless your G-code is regularly overshooting the limits of your machine, you really only need three. Those would be for Xmin, Ymin and Zmin – that is, the origin in X and Y and the highest point your toolhead will go (remember that on the 3018, the toolhead goes down, so Zmax is below the surface of your workpiece but is mechanically limited by the amount of travel the toolhead will travel).

Let's start with the end-stops themselves. These are nothing more than ordinary tap switches that are normally open, and once the switch is actuated, close the circuit. Here's an example switch (without wires):

Figure 6.1 – Simple end-stop switch

Take close notice of the indications on the terminals. You can see **NC**, **NO**, and **C** on the body of the switch. More about this later.

You can buy these switches just about anywhere. I've also sometimes harvested them out of various machines I take apart. You only need two wires and a two-wire plug for each switch. Mounting these switches is simple and can be done directly onto the frame of the machine. There is another type of mount that puts the switch on a mount that fits around the rods/rails of the machine. Some of them are a press-fit, and others can be secured with a screw to hold them tightly onto the rods. That particular setup can take away a few millimeters of work area though, which is why I prefer to mount my end-stops on the frame.

Before you mount your switches, you will need to decide where you want your toolhead to home to. I like to home mine to the rear right corner (when facing the machine). Therefore, I place my X and Y switches on the right side and front of the machine respectively. Remember that when you home, the machine's controller will move the axes until a switch is tripped, so when I put my switches in that rear right corner, I have to tell my control software that the origin is there so that the toolhead goes in that direction when I send a home command in G-code. Here is a photograph of my X end-stop:

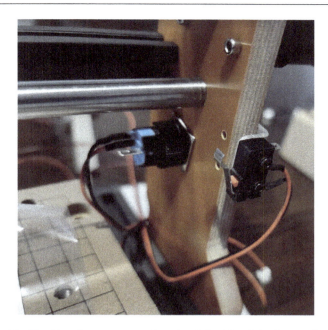

Figure 6.2 – The X end-stop mounted on the right side of the frame of my 3018

Note the type of switch and the pins to which the wires are soldered. The switch is mounted using a small metal bracket that you can make out of any piece of sheet, aluminum, or L-shaped bracket you can find. In my case, the bracket came with the kit for this particular machine.

The following is a photo of my Y end-stop:

Figure 6.3 – The Y end-stop mounted on the front of the frame of my 3018

Note the bracket that came with the kit. Depending on your frame, you may also be able to mount the switch right on it (for example, if you have a frame made from extrusions). What is important is that the table can trip the switch when it reaches it.

Of course, there are many ways to install these end-stops. What you use depends on your machine and how the frame is put together. Here are some example mounts that I 3D-printed from various sources. You don't need to look far for these if you can 3Dprint things: almost everything I have found came from Thingiverse. You can even re-purpose components as well: for example, the blue mount in the following figure is from a Delta 3D printer:

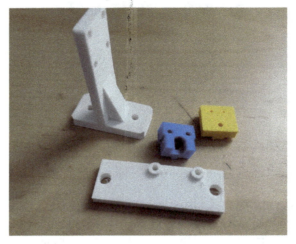

Figure 6.4 – Some example mounts for end-stop switches

You can also use different types of switches to suit your purposes. For example, on the X-axis, you can use the following switch and mount it on a suitable surface:

Figure 6.5 – An example of a switch mounted on its own PCB. Note
the switch orientation compared to the other switches

Here's another switch mounted on a multi-purpose mount that can be used in X and Y:

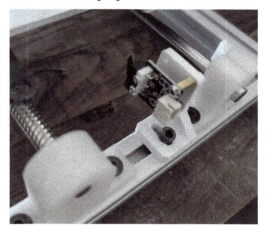

Figure 6.6 – This type of switch is used on Ender-3 printers – here, it is installed for
the Y-axis on one f my machines that uses 2020 extrusions for its frame

This switch is very common on Ender-3-type printers. Like the switch in *Figure 6.5*, it is a 3-pin switch, and you just have to use the two wires that matter for our purposes. More on this in a bit.

Here is the same mount with the same type of switch mounted for the X-axis, again on the same machine that primarily is built using 2020 aluminum extrusions:

Figure 6.7 – Same switch, same mount, different axis

Finally, here is the Zmin (top limit) switch installed directly on the X carriage:

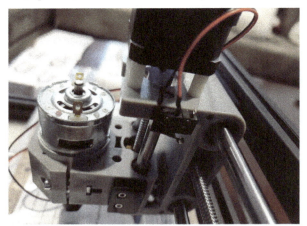

Figure 6.8 – Same switch as X and Y above, but mounted on Z. Note
the wires and what terminals on the switch are used

There are two small holes drilled into the plastic that are enough to hold the switch. I also added a little piece of acrylic to actuate the switch when the Z carriage rises high enough. I have two more holes drilled on the lower part of the X carriage to mount another limit switch that prevents the end-mill bit from digging into the worktable.

Now we need to discuss the wiring of your switches. If you have an ordinary tap switch like the one in *Figure 6.8*, you only need to have two wires coming from the switch and going to the controller board. The following figure is a common board for CNC with a lot of options, so I am going to use it as an example. This board requires end-stops with two wires:

Figure 6.9 – My 3018's controller. Note the 2 pin connectors along the
bottom. This is where my end-stops are connected

Most mechanical switches have their terminals labeled **NC, C**, and **NO**. **NC** stands for **Normally Closed**, **C** stands for **Common**, and **NO** stands for **Normally Open**. You are interested in the NC and C terminals. Sometimes, these are also labeled **1, 3**, and **2** (note the sequence). On my 3018, all my switches are wired to the 1 and 3 terminals, but if you have NC, C and NO labels, you are interested in the NC and C terminals. NC goes to Sig (Signal) on the board, while C goes to Gnd or Ground. Three-wire end-stops will have a Sig, Gnd, and VCC (or V+). You are again interested in the Sig and Gnd terminals on the switch here. Teaching you how to solder is out of the scope of this book, but your objective is to connect a pair of wires to those two terminals as illustrated in *Figure 6.9*. If your board has three-pin end-stop terminals, then you can use the wire that came with the switch. You can also use female-female jumper wires. Of course, you can also just buy pre-wired switches. The following table explains the various terminals or pins on most of these small microswitches:

Terminal/pin label	Purpose	What to do with it
C	Common	This is ground (like the negative terminal on most electrical devices). I usually attach a black wire here to tell me that this is the ground wire so I don't have to guess or look too hard to know what it is.
NC	Normally Closed	When the switch is "closed", i.e., pressed, electrical energy is interrupted and so there is no flow of power through the switch. When an end-stop is hit, I want this to happen because that interrupts movement on that axis. When I add wiring to this terminal, I use wire with red shielding, denoting a "live" connection (i.e., power flows through it all the time).
NO	Normally Open	When the switch is "closed", i.e., pressed, electrical energy is able to flow through the switch. Until then, no power flows through the switch at all. I do NOT want to use this contact.

Let's now look at one other type of switch – one that will halt all operations in the event a critical error occurs.

Emergency stop switch

The purpose of this switch is to bring everything to an immediate stop in the event your toolhead is about to damage something or go beyond its cutting borders (or any other emergency). It doesn't matter what sort of switch it is, but it must have a way to positively break the circuit. I personally like big red push buttons such as the one you see here on my machine:

Figure 6.10 – My E-Stop switch, mounted right on my 3018's frame

You can find these on Amazon or similar sources for very little. They will have three terminals but are wired with two wires. My controller board has a two-pin terminal for this switch, but if yours does not, wire it to your connector to the board (the live, i.e., positive wire). Do not wire it directly to the plug to the wall as that would be dangerous. Install the switch somewhere easily accessible from any angle on the board. The reason I like the big red button is that you cannot reset the machine back to operation without expressly twisting the red cap to release the switch. It is very difficult to accidentally turn the machine back on. You don't have to use a switch like this, as a normal toggle switch will do fine as well, but I would select one intended for heavy-duty use, so that there is no chance of accidentally hitting it.

Let's extend the functionality even further now, using a probe on the Z-axis to compensate for any variations in the end-mill length and how the workpiece is positioned.

Installing a Z-probe

Irrespective of your end-stops, your end mills are going to be of different lengths and if you replace the spindle with the drag knife or pen holder, then the starting point for machining (i.e., where you set Z = 0) will change. This is why I like to have some sort of Z-probe to set the top limit of my Z-axis with every job. That's the job of the Z-probe. You can buy these just about anywhere, and they look something like the following:

Figure 6.11 – A Z-probe just as it comes out of the packaging

On my controller, there is a connector for this, and there should be one for yours as well. This is where your connector will go. My Z-probe did not come with a connector, but you can make a suitable connector using a jumper wire or a suitable two-pin connector for your board. I find that using a probe is my preferred way to "zero" my end mill and toolhead. The procedure is very simple after you have wired in a connector and plugged it into the controller board:

1. Insert your bit into the spindle and lightly tighten the collet.

2. Lower the Z-axis to the maximum point it can be and loosen the collet allowing the bit to slide down and touch the worktable (which at this point will likely be your waste board). Hand-tighten the collet.

3. Measure the height of the probe with a caliper.

4. Plug the 3018 into your computer and launch UGS.

5. Press **Reset Zero** on UGS and raise the toolhead by an amount close to the height of the probe. My probe's height is just slightly less than 20 mm and appears to be somewhere around 19.5 mm.

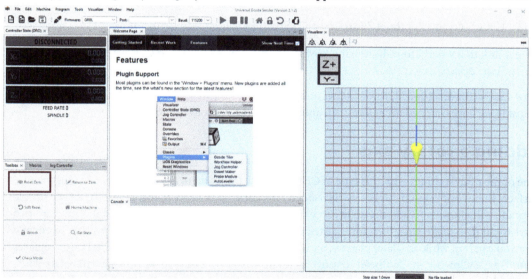

Figure 6.12 – The Reset Zero button on UGS

6. Now raise the toolhead in small increments of 0.1 mm and check that the probe will just about slide under your end-mill bit, possibly lightly touching the surface of the metal plate on the probe. The height you read on UGS should be the height of the tip of your end mill above your waste board. Compare the two numbers and if they are fairly close, then you can move forward. Expect a difference of 0.2 mm or so between the two measures.

7. Plug your probe into the designated header on your controller board. On my board, this is indicated in the schematic for the board. Check your controller as to where you should plug it in.

8. Raise the toolhead so that you have some room to work with, mount your workpiece, and place the probe under your end mill.

9. Clip the alligator clip's jaws onto the end mill's shank.

10. Start up UGS, connect to your machine. Once connected, go to your probe menu (**Window | Plugins | Probe Module**).

11. Click **Settings** and generally set the units to millimeters, work coordinates to **G54**, end-mill diameter at **10**, fast find rate at **100**, slow measure rate at **10**, and retract amount to **1**. This will allow the toolhead to descend quickly first to touch the probe, then rise 1 mm and come down much slower to touch a second time for a very accurate measurement.

12. Now click the Z menu and enter the probe height (labeled as thickness) as measured by the calipers and set the probe distance/direction to -**20**.

13. Now click on **Initiate Probe**. This will lower the toolhead until it touches the probe plate twice (first quickly, then slower). Once done, your machine has been zeroed. You can now remove the probe.

14. Click " (do not click **Reset Zero**). This will lower the toolhead to touch the surface of your workpiece.

Here a link to a video I found, showing how this is done using UGS and Candle: `https://www.youtube.com/watch?v=3xgIS_IEX-Y`.

Now that we have added all the switches and major controls, we can turn our attention to another upgrade we have touched on before – the fourth axis, which for us is a rotary one.

Adding a rotary axis

One of the neatest things with this type of CNC machine (and others like it) is to cut or carve a round surface. In our applications so far, we have been operating on the X-Y plane. However, a very simple upgrade will allow us to add the ability to work around a surface instead of just on it. There are two ways to do this: make your own rotary axis, or just buy one. In this chapter, we are going to cover a unit you can buy, while in a later chapter, we will discuss fabricating your own. The rotary axis replaces your Y-axis, allowing the toolhead to move in the X direction, but your workpiece will rotate while that is happening. The nice thing about all this is that you can carve/cut with the spindle, engrave and cut with the laser, and draw with the pen. A rotary axis is therefore a very useful upgrade. In this section, let's quickly discuss the setup and then look at the impact on the G-code. We will do a deeper dive into working with a rotating workpiece once we have had a chance to work on some projects.

Here's the unit I purchased. It was inexpensive and I liked that it was all metal. Note that it has its own stepper motor. Take a look at `https://www.comgrow.com/products/comgrow-laser-rotary-roller-engraving-module` to see how it sits on your 3018 and some examples of objects it can engrave.

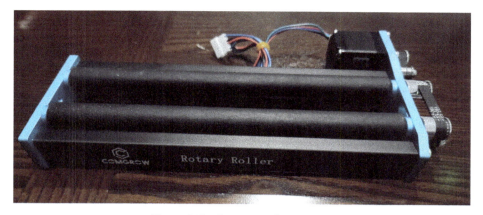

Figure 6.13 – Rotary axis for my 3018

This unit is very compact. To connect it, you simply unplug the Y-axis motor from your board and plug in the motor wire for the rotary axis. This particular unit comes with the ability to adjust the rollers to accommodate narrower or wider objects. This add-on is useful for objects with work surfaces that are constant, but if I wanted to work on something that has a radius at one end different from that at the other (such as a wine glass, coffee cup with a handle, or many other forms of glasses with mouths wider than their bases), I would need to tilt the object to keep its distance from the end mill the same across the X-axis. The unit shown in *Figure 6.13* is not suitable for this. Our project later in this book involves making a rotary axis that accommodates irregularly shaped cylindrical objects.

If you purchased one of these, you will likely see instructions to adjust the steps per mm for your Y-axis before you use it. You might also be instructed to change the acceleration value. Keep note of what these are prior to the changes so you can change them back when you are done using the rotary axis. You can do this by opening UGS, connecting to your 3018, then opening the console window and typing $$ in the command box. That will show you a full list of the settings in the console window. You can see where the console window is in *Figure 6.12* (center panel, towards the bottom).

Depending on the unit you acquire, you may also have to calculate these values based on the microsteps for the motor, the number of steps per revolution, and the number of mm traveled per revolution. The formula for steps per mm is as follows:

Steps/mm = steps per revolution x microsteps per step/mm per revolution

Follow the instructions provided to identify what those settings should be. You can set the values using GRBL using $101 (Y-axis steps/mm) and $121 for the Y-axis acceleration.

To change the values, issue the commands like this:

```
$101=65
```

This will tell UGS that your machine's Y steps/mm setting is now 65.

Your controller doesn't know the difference whether you have the rotary axis on or not, so once you have the steps/mm dialed in you'll be able to operate on your cylindrical surface as if it were flat and your 2D or 3D "flat" design rendered in G-code will instead be rendered on your cylindrical workpiece. This is particularly fun with a laser, because you can now inscribe/engrave your coffee cups and glasses with intricate designs that would not otherwise have been possible. As a final note, I wanted to point you to this video to see how you might calibrate your axes should you find that your steps/mm are not what they should be: `https://www.youtube.com/watch?v=nGwUf8uNWBU`

Plotters and drag knives

Upgrading the 3018 to be able to make precise drawings or cut vinyl, cardboard, and paper mechanically is the least complicated modification you can make to your 3018. For the most part, you are replacing the spindle (temporarily) with something to hold a pen/sharpie or a blade. The rest is controlled by the software and G-code. Consider: If you have no traversal in Z other than to raise or lower the toolhead, your drawing is rendered entirely in 2D (no depth to worry about), so it is reasonable to think that your 3018 can draw and cut using the same software. You would just have to work on the process of raising and lowering the toolhead where appropriate.

While these modifications are certainly simple enough and absolutely reversible, you should know that the machine will appear to operate slowly. This is because the 3018s use leadscrews in all axes and those are meant to move a heavy toolhead at a reasonable rate for cutting. In the case of these upgrades, you are still bound by the traversal speeds, which means that unlike belt-driven plotters and cutters (such as the CriCut, for example), your drawings will be slow to render and you might in fact find it to be more useful to modify a 3D printer or fabricate a specific fit-for-purpose machine (which we will cover later in this book). A fit-for-purpose machine will use much the same hardware as the 3018, but movement is done via belts and pulleys on linear rails or bearings.

We will deal with the changes needed to implement in hardware later in this chapter, but I think by now you should know enough about generating G-code for a drawing. Of course, the thickness of your material will be zero. In both cases, you will be removing and disconnecting the spindle and then inserting something to hold the pen or cutting blade. Let's start with the drag knife.

The first thing we need is something to hold the blade. You can buy a suitable holder from a variety of vendors. Here's an example I bought on Amazon for about $15:

Figure 6.14 – Drag knife holder and example blades

The blades in the preceding figure are 30/45/60-degree blades. The numbers denote the angle of the blade. The set I picked up online comes with a holder. The red-tipped blades are the 30-degree blades, the yellow ones are 45-degree, and the blue ones are 60-degree blades.

While this holder unit is what we need, we still need a way to mount it on the toolhead. There are a number of these featured on Thingiverse, like this one: `https://www.thingiverse.com/thing:4014528`. It's a two-part unit, but you'd have to 3D print it. I like this design because it could also double as a pen/sharpie holder (you can remove the drag knife mount and just insert a pen in its place). On YouTube, there are videos showing how to fabricate one using parts from a DIY hardware store, but for our purposes, a printed unit is fine. While surfing eBay, I even found a mount for sale to hold a pen, which could also be used to hold this mount if you prefer to buy rather than build your own.

Once you have the drag knife holder mounted on your 3018, you are pretty much ready to go. Here's what it looks like mounted on your 3018. Note that the blade is barely out of the holder but just enough to cut whatever it gets dragged over:

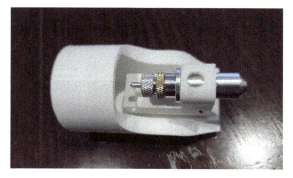

Figure 6.15 – Drag knife and holder and mount for the 3018 toolhead

The final element is G-code, so we move to software. We've already seen how we can take an STL or even an image, import it into something like TinkerCAD, export it back out as a vector graphic, and then from there go to G-code. I used various tools in the previous chapter (including FileStar and InkScape) to convert the file. With no thickness, your 3018 will raise and lower the toolhead where appropriate and cut through whatever material you put underneath.

To get cutting, simply zero your toolhead to the surface you are cutting (i.e., lower it through the control software such as UGS) until the blade is just about in contact with the material you are cutting and then go ahead and send the G-code over to the machine. You know have a fully operational vinyl cutter.

The plotter modification is not much different. Just as before, you need something to hold the pen or sharpie (or whatever you are using to draw) and the rest of it is exactly the same. As I have already mentioned, you can purchase a pen holder on eBay, learn how you make your own out of readily available materials on YouTube, or simply 3D-print one (my preferred approach). Here's an example of one on Thingiverse: `https://www.thingiverse.com/thing:4114678`. Another is this one, `https://www.thingiverse.com/thing:3533806`, which is a bit more substantial. Both fit right into the toolhead and are secured by the same mechanism that holds the spindle in place. You can see how simple it is to fabricate something like this and modifying these designs to hold different-sized pens/pencils is fairly straightforward.

Summary

So far, we've covered mostly mechanical upgrades to our 3018. There are more that we should cover, but they are very quickly becoming projects of their own. The first one we want to cover next is adding a laser and discuss what the impacts are on our operations when we do. The other project is to review how we can further improve our rotary axis. However, from our work in this chapter, we now have a machine that can home to the same place every time, a way to accurately zero our Z-axis, and a way to stop the machine in an emergency, and our machine can now double as a plotter and drag knife, albeit a slow one.

You will have noticed the utility of having access to a 3D printer due to my by-now multiple references to Thingiverse and similar sites. This is because there is a wealth of information there that you can lean on. I try very hard to not reinvent the wheel, especially with designs. I might modify a design, but if I can use it as is, so much the better. Definitely keep that in mind when you are looking for something specific.

7
Enclosures

If you are anything like me, the one thing you don't relish is cleaning up after you are done with your project. I always want my machine to be right where and how I left it, ready for the next task, but I also know that an organized workbench also means better focus on my task at hand. And, with CNC, we really should attend to the debris that results from our efforts. This debris can contaminate surfaces we are working on or with, and inhaling any of the dust (or fumes) is not particularly healthy. As a result, we really should consider putting our machine in an enclosure to at least control where the debris and dust go and limit how much cleanup we need to do. This chapter deals with deciding on and either assembling or buying an enclosure for the 3018. I like to have enclosures for many of my machines. A number of my 3D printers live under a hood to allow heat to be retained and noise to be dampened; still others have a cube frame permitting all the working parts to be enclosed with flat panels. My larger 10W laser lives in an all-metal enclosure not just to contain fumes and dust but also fire, which has happened on one occasion due to a piece of material that came loose mid-cut. Your particular enclosure should fit your needs – if you are not going to use a laser, you will not need polycarbonate panels, and if you are not using oily lubricants, you may be fine with fabric-based enclosures.

We will focus on identifying what your enclosure should be capable of, and identify the parameters around the build-or-buy decision:

- Considerations if your 3018 will be both a CNC and laser cutter
- Fire/heat needs and requirements
- Ventilation requirements and whether you plan to use a vacuum to pick up dust and debris
- Ease of cleaning

We will learn learning what materials we can use to fabricate an enclosure based on the aforementioned requirements as well as be able to evaluate what works for us should we choose to purchase one. As part of this conversation, we will also go over some enclosure designs and how we might go about fabricating a custom one.

Panels, cleaning, and access

At first glance, the three elements in this heading might seem unrelated, but in reality, they absolutely are. You want to have panels that open easily and provide maximum access to your work area as well as the controls on the machine itself. Of course, removable panels or a way to remove the entire enclosure for cleaning also means you have a way to control the spread of any debris that collects with each cutting job. Panels provide the means for ventilation and cooling you need to drive harmful fumes away from your machine and work area or let in air to keep your workpiece or toolhead cool. Also, with panels, if you are using a laser, make sure that you can safely view your machine in action without jeopardizing your eyesight. These same panels may also provide you with access to your emergency stop switch, power supply, and USB cables, and also confine any sparks or flames (should your workpiece ignite).

What you do for your enclosure panels depends on how your enclosure is set up as well as what sort of work you do with your 3018. An enclosure, for example, does not have any value if you are using the 3018 as a plotter or a drag knife. Let us then turn our attention to the sort of enclosures you might want to have. The following two sections will give an idea about the two different types of enclosures.

Removable and non-integral enclosures

Essentially, an enclosure is a box that sits around your machine. It has no bottom panel so that you can pick it up and place it over your 3018 much as you might cover something with a cardboard box. A lot of pre-made enclosures are like this. They are essentially a frame with some transparent panels, one of which may operate as a door, that you lower over your 3018 when in use.

Cleanup is fairly simple since the entire enclosure is removable. Once done, simply lift up the enclosure and clean it with rags, towels, or a shop vacuum as necessary – both the interior of the enclosure as well as on, around, and under the 3018. There are plenty of vendors on the internet who will sell you one of these for less than $100 and assembly is trivial. Enclosures like these have their advantages and downsides.

The following are some of the pros:

- Simple to build and use
- Does not require any modification to your machine
- Can be shared between multiple machines (if you are like me and have several 3018s)
- Does not need doors and not all panels have to be transparent
- Can be made from simple, lightweight materials (e.g., some use fabric-covered frames like this: https://www.amazon.com/dp/B0B1ZWMXXB)

The following are some of the cons:

- Integral features of your 3018 may not be reachable without removing/opening the enclosure (e.g., the emergency stop switch may not be reachable, or the LCD controller ribbon cable may not reach through or under the enclosure frame).

- Power and USB cabling might interfere with the placement of the enclosure. This is something to watch out for, especially with enclosures you buy online. You must have at least one reasonably sized hole to allow both USB and power cables to come through. Here is an example: `https://www.amazon.com/Enclosure-Protection-Dustproof-Reduction-Compatible/dp/B0B977K8NP`.

- If you have a dust shoe and vacuum hose attached to your machine, or cooling/air assist hoses, a non-integral enclosure may make getting to your machine between jobs a chore.

By the time you are done, you might find yourself permanently installing components into your non-integral enclosure just to reduce the hassle of having to connect/disconnect components every time you want to get to your machine.

Figure 7.1 – A tube-and-fabric enclosure for my 3D printer – easily usable for my 3018

Having seen some ideas of flexible enclosures, we will next look at those that are less temporary and can be part and parcel of your machine.

Integrated and permanent enclosures

Integrated enclosures offer some unique benefits and make your machine seem more *professional*. Typically, I like enclosures where the machine can be bolted in place so that both the enclosure and machine move as one unit around the shop. I look for enclosures that have solid walls and the ability to support peripheral items on the outside. For example, I like to have the emergency stop switch outside so that if I need to halt any operation, I don't have to reach into the box in any way. I also like to separate the electronics, dust collection, lubrication, and air assist systems so they can be controlled from the outside – again, so that I don't have to put my hand inside while there is debris flying around. Here is an example of an integrated enclosure around my 10W laser:

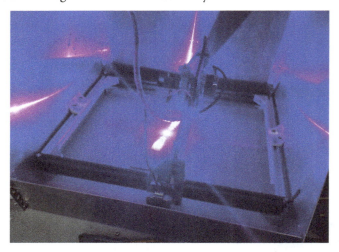

Figure 7.2 – My 10W laser with its all-metal enclosure (and the laser firing)

Figure 7.3 – The exterior of my all-metal enclosure for my laser. Note the placement of all the electronics and controls at the top of the enclosure

This enclosure started out being designed for a 3D printer. I found the plans for it somewhere on the internet and I think the page where it came from is long gone. However, I had downloaded the files and kept them on my internal network. The original design was meant to be cut from plywood or **medium-density fiberboard (MDF)**, but I decided to have it cut from aluminum. In the process, I miscalculated how sturdy the structure would be with a 3-mm aluminum sheet so I added some angle aluminum in the corners to make a very solid structure. 3D-printed hinges (you could use small cabinet hinges instead if you were to do something like this), a door handle (an ordinary drawer handle would have worked fine too), and a pair of brackets left over from some other project completed the assembly. The window is a polycarbonate acrylic to block laser light (it's a blue laser) that is also easily covered but can accommodate any laser toolhead (i.e., a green or red laser). The reason I went with a metal enclosure is because at one point, my machine jammed and the laser ignited the workpiece and the resulting fire melted a lot of the plastic parts. This way, the enclosure might heat up, but it won't burn. The plastic parts are small enough to burn away quickly and any damage will be localized inside the interior of the box. That is not to say your enclosures will catch fire too, but for me, I typically adopt a once-bitten-twice-shy policy to remove any chance of something like this happening again. Here's a picture of the post-fire result of my machine. I had to cut down the frame as a result to fit it into the enclosure.

Figure 7.4 – Melted plastic and damaged wheels and frame from an ignited piece of Depron being machined by my laser

Let's review some of the pros and cons of a fully integrated enclosure.

The following are some of the pros:

- No need to get inside the enclosure to stop or control the machine so long as any controls and electronics beyond the toolhead are mounted on the outside. This is the primary benefit of an integral enclosure.

- It is also easier to route hoses and other plumbing so there is no need to constantly carry all that apparatus around with the machine.

- The enclosure and machine move as one unit and there is no need to adjust or reset the machine's position inside if it is relocated.

- Integrated enclosures *may* be sturdier and have better fire protection, although this is not necessarily the rule. A non-integral enclosure can be built just as sturdy, although typically, that is not the case if one is bought off the shelf.

The following are some of the cons:

- The enclosure is married to the machine. Any upgrade to the size, toolhead, or some other aspect of the machine may require a modification or replacement of the enclosure.

- There is no machine-to-machine reuse of the enclosure.

- Extensive maintenance will require you to remove the machine from the enclosure and possibly disassemble the enclosure itself.

Your choice of enclosure and its features entirely depend on your use case and everyone will have different requirements. Because I have multiple machines of varying sizes with differing configurations, I tend to flip between integral and non-integral enclosures.

In the next section, we will look at the decision of whether to custom-make your own enclosure or buy one commercially, and if you do choose to make it, what decisions need to be made around materials and designs.

Build versus buy, materials, and designs

There are a lot of options available to you depending on how confident you are working with easily available materials. If I had to make design choices for any enclosure, I prefer to have the following:

- Front-mounted doors, with a preference for doors that open up and away (i.e., with the hinges mounted on the top).

- A clamshell-type design, which allows me to have complete access to all sides of the machine with the enclosure open. This is one of the designs I have listed previously.

- Plenty of openings for wires and hoses. If it is an integral enclosure, I either need enough real estate on the top or side to mount any controllers or long-enough cabling for the controller to mount to the door or an inset at the front. This is what I ended up doing for my laser. The controllers are all on the outside, as is any hardware to job the machine (e.g., an LCD controller).

A cursory look online yields a variety of available units that you can buy. There is just as much variety for the enclosures you can build yourself. As a maker, I would violate my own ethos if I purchased an enclosure, but if my requirements are fairly simple, then a store-bought unit might suffice. Keep in mind that, given the price of a 3018 machine, an enclosure that costs half as much as the machine might not be money so well spent (although that is just my opinion), as I would rather spend that money buying a bigger and more powerful unit instead, or some fun upgrades. Still, if you decide to purchase a ready-made one, I would avoid the fabric, tent-like units and consider one that insulates not just from debris but also from noise. Some enclosures have sound-absorbing material in the walls that can quieten down what you hear coming from the machine. Just remember that if you have walls with foam-based sound insulation, vacuuming dust and debris out of them will be tedious.

If your requirements are stringent and a ready-made enclosure is out of your price range, you can build one out of materials easily found at your local big-box DIY store. For example, the structure can be made of MDF, assembling the walls using T-nuts, glue, or ordinary screws. You needn't worry about perfect seals because you can add a foam door or window seal to prevent anything from escaping the enclosure. Cut out some nice side windows and insert some clear acrylic sheeting mounted from the inside for a clean look. If you want to ensure that less sound escapes, limit the windows to just the door and keep its size to a minimum.

You can also opt to have your enclosure parts cut professionally by a number of services. The metal box you see in the preceding images was cut by Xometry, but I have also gone to other companies including Ponoko. These services will provide the material, cut it with a waterjet or laser, and ship it directly to you, ready to assemble. You can fashion your enclosure out of thick acrylic, aluminum, plywood, or any other sheet material sturdy enough.

If you prefer to make your enclosure non-integral, you can choose to not give it a bottom or floor. This allows you to simply lower the enclosure over your machine. As I have mentioned already, I prefer clamshell-type designs (e.g., this one on Thingiverse I find particularly interesting: `https://www.thingiverse.com/thing:4265505/files` – note the sound insulation shown in the image), but a bottomless enclosure is fairly simple and you can still get to all parts of your 3018 by just lifting it up and away (although I would never do that while it is running).

A quick search through YouTube will yield several designs that you can duplicate for yourself. One example is this, made from plywood, although it is not a clamshell design: `https://www.youtube.com/watch?v=Jc-nVEshfCc&t=85s`. Notice the foam sound insulation and the door hinged from above, much the same as my metal box. I much prefer rubber over foam sound insulation, and rather than the thinner plywood, consider using thicker material (plywood or MDF). The box may be heavier, but the machine will be quieter. Ultimately, what the enclosure looks like is entirely determined by you.

Summary

Together, in this chapter, we have learned what we might expect from our machine's enclosure and considered whether it is better to build one or just buy something simple. We also looked at some fabrication methods and saw a design you can build yourself out of commonly available materials. We also discussed the differences between integral and non-integral enclosures, noting the benefits and downsides of each.

We also briefly touched upon third-party fabrication services, which may be suited for larger projects or to make parts you are not able to make yourself yet. Later in the book, we will discuss larger machines that you can have the 3018 help fabricate. I will, in particular, highlight my BumbleBee combination CNC and laser cutter, which you can then use as a model for your own projects.

In the next chapter, we will look at a project to make a different kind of CNC machine – a plotter/laser cutter. Even though we discussed making the 3018 do this for you anyway in *Chapter 6, Upgrading Your CNC Machine,* you will find that these machines, which are largely belt driven, are faster and can be customized to have larger work areas. They use the same concepts and the same software (GRBL). Why build a plotter when you have, say, a desktop printer? Because with a plotter, your renderings will be less susceptible to the limitations of the printer (paper size) and, of course, these plotters are lighter, simpler, and just way more fun.

8

Project: Building a CNC Laser Cutter and a Plotter

As we have already seen, CNC machine use cases involve more than just shaping workpieces with a spindle. We can also convert our 3018 machines into laser cutters, plotters, and drag knives. However, because the 3018 uses leadscrews to handle the rigidity requirements of working on blocks of material, it is generally slow when used as a laser cutter. Most commercially available machines use belt-driven systems for motion. This is because the toolhead typically does not experience a load from the workpiece.

In this chapter, we will build two machines. The first one, a laser cutter, will be built using an adaptation of a commonly used frame, which we will adapt from a plotter design so that we only use off-the-shelf parts. While it is possible to buy a unit commercially, building one will give you a better appreciation for the mechanics of a belt-driven motion system. It will also allow you to customize the machine to your specific needs as time goes by and your skills evolve.

The second machine we will build is a plotter based on the same architecture. You can build exactly the same machine with the same hardware with just one variation – the toolhead – and have a high-speed plotter capable of rendering drawings on any flat surface your plotter pen works on. With some firmware tweaking, it is also possible to use the same machine for both purposes.

Our objective here is to focus entirely on the mechanics of the machine since we already know how the electronics and software work. As this is a chapter that deals primarily with project work, you will do the following:

- Gain insight into different motion systems
- Gain familiarity with components that you can use to improve your 3018 and build more complex machines in the future
- Learn how to develop purpose-built CNC machines and extend your fabrication capabilities

Laser cutter/engraver

The frame we are going to be building is based on carriages that use V-wheels. These are commonly available online from a variety of sources. Let's begin with a brief discussion of the general design. Like the plotter, the laser cutter uses a cantilever arm for the Y axis on which the toolhead moves. The entire arm moves in the X axis and the toolhead can be any desired laser head you desire. You can use exactly the same type of board you have on your 3018 as the motherboard for this machine and you can flash GRBL on it, just like the 3018. In addition, you can also attach an LCD controller to operate it offline (i.e., without a computer). You can also attach some other commonly available boards such as the MKS DLC 2.0, for which you can acquire a TFT controller flashed with firmware specific to laser cutter operations.

Here's a bill of materials for the laser cutter frame:

Item Description	Quantity	Source/Link
2040 V-slot extrusion 400 mm (X axis)	1	https://www.amazon.com/dp/B09LQJ3PJ5 (you only need one extrusion from this set of four, but if you are building the plotter too, you will consume three of these and have an extra left over)
2040 extrusion (standard or V-slot) 100 mm	2	You can cut up one of the pieces in the preceding set since you get four extrusions of 400 mm each
2020 V-slot extrusion 500 mm (Y axis)	1	https://www.amazon.com/dp/B09DTL7G6X (again, you get four pieces, so you will need a second one of these for the plotter)
V-wheel carriage and wheels for X axis	1	https://www.amazon.com/dp/B0B99WTBSY (you will only need four of the six wheels this comes with)
V-wheel carriage and wheels for Y axis	1	https://www.amazon.com/dp/B0919HJ21D
X-axis belt tensioner with GT2 belt pulley	1	https://www.amazon.com/dp/B08QZ6S23J
Y-axis belt tensioner with GT2 belt pully	1	https://www.amazon.com/dp/B07XL41QNH
NEMA 17 stepper motors	2	https://www.amazon.com/dp/B08Z7L2P6S (most NEMA 17 motors will work)
Y-axis stepper motor mount	1	https://www.amazon.com/dp/B08CDWVHRW
X-axis motor mount	1	https://www.amazon.com/dp/B07Y544MZZ

GT-2 pulley, 5 mm bore, 20 teeth	2	https://www.amazon.com/GT2-Creality-Ender-3-Printer-Stepper/dp/B088WB8D7W (the set comes with more; they always come in handy)
G-2 timing belt	1	https://www.amazon.com/dp/B07BRKZGMS
2020 brackets	4	https://www.amazon.com/Aifeier-Connectors-Aluminum-Extrusion-Accessories/dp/B0BW5RGNS1 (the extra four you have will be useful for the plotter)
M3 screws to mount the motors M5 screws and T-slot nuts to mount brackets and parts to extrusions	Various	

Assembly is straightforward and should be fairly intuitive. Here's an annotated photo showing the fully assembled frame with all the mechanical components installed:

Figure 8.1 – Fully assembled laser cutter frame

The 2040 extrusion (wider, shorter extrusion material) is the *X* axis, and the 2020 extrusion is the *Y* axis (longer, narrower extrusion). In the following steps, I will show you close-ups of the individual extrusions, and having assembled a 3018 machine, you should have no issue following along. We will start with the *Y* axis.

Install the motor mount on the *Y-axis* 2020 extrusion. You can see, in *Figure 8.2*, that I have fastened the plate to the extrusion and then mounted the wheels before sliding it onto the 2040 extrusion. This is no different than when you installed motors on the extrusions of your 3018 machine; the only difference here is the orientation of the motor. I also have the motor mount installed with the stepper motor in place – in *Figure 8.1*, this is in the middle of the photo, on the left.

Figure 8.2 – The Y-ais carriage mounted and slid onto the 2040 extrusion

From here, you can slide the assembly onto the 2040 extrusion. At one end of the 2040 extrusion, mount the 2040 belt tensioner, as depicted in the following figure:

Figure 8.3 – The X-axis belt tensioner mounted on one end of the 2040 extrusion

Attach two pieces of the 100 mm extrusions on each end of the larger 2040 extrusion. Use the brackets (two per side) to make what amounts to a long "U" shape, lining up the edge of the 400 mm 2040 extrusion with the surface of the 10 mm extrusions:

Figure 8.4 – 100mm extrusion "feet" mounted on the end of the 400 mm 2040 extrusion. Note how I have put the brackets on the other side to hide them

The brackets to make this work are on the "inside" of the assembly to give a clean surface on either side of the machine, as depicted in the following figure:

Figure 8.5 – 100 mm extrusion attached with brackets

You will need to tap the 2040's holes on the side where you don't have the tensioner. Use an M5 tap for this, as should be evident here. On this side of the extrusion, install the motor mount and the X stepper motor. You can see this in the following figure:

Figure 8.6 – Motor mounted to the 2040 extrusion

In the next figure, you can see another view of the motor mount and motor. Here, I have installed one of the pulleys and run the belt through the extrusion. You can see how it is bolted onto the 2040 and the machine's foot:

Figure 8.7 – Another view of the motor mount for the X axis

Looking at *Figure 8.1*, you can see where both ends of the belt have been attached to the *X-a*xis carriage. Generally, I like to clamp the end of the belt in a loop using two zip ties. The teeth of the belt on either side of the loop lock against each other and provide a firm lock.

It's fairly simple now to assemble the *Y* axis in a very similar fashion. At this point, you just need to add the tensioner on the other end of the *Y-a*xis extrusion. This is the 500 mm 2020 extrusion, as shown in the following figure:

Figure 8.8 – X-axis belt tensioner in place at the far end of the 2020 extrusion

Unlike the 2040 extrusion, I have passed the belt on the *side* of the extrusion because the *Y-a*xis carriage is vertically oriented.

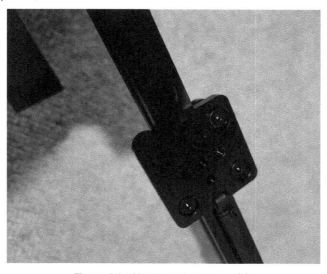

Figure 8.9 – Y-axis carriage assembly

The *Y-a*xis carriage assembly is straightforward. With the three tapped holes on the carriage, you can mount any toolhead that suits your purposes, which, for us, will be the laser toolhead. Depending on the laser, you will have to fabricate or adapt a mount. That should be straightforward.

Here's a figure showing the mostly assembled machine without the belts installed. Note that I haven't put the GT2 pulley on yet in this figure:

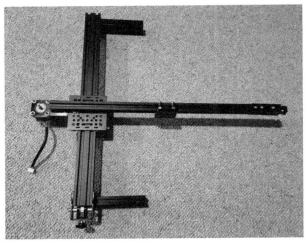

Figure 8.10 – Top view of the mostly assembled machine

Note the orientation of the *Y-a*xis carriage. This is important because it's very easy to forget and find yourself having to disassemble something to make corrections. For further reference, the following figure shows the bottom of the machine:

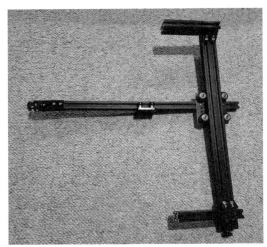

Figure 8.11 – Bottom view of the mostly assembled machine

In the preceding figure, I had not yet installed the *Y-ax*is motor mount, but it obviously goes on the end of the 2020 extrusion.

At this point, you can loop and lock the belt, as seen in *Figure 8.10*, and you have the mechanics largely done for the laser cutter. Here's a figure showing the completed machine without the laser toolhead installed. You can see the belts and pulleys in this figure.

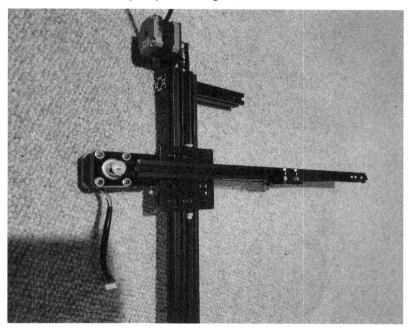

Figure 8.12 – Machine fully assembled

Once the belts are in place, use the tensioners to tighten them and your machine now only needs the toolhead and electronics installed.

Installing the laser toolhead entirely depends on the sort of laser you acquire. If you purchase a laser that has a fixed-focus lens, you will need a mount for the laser that allows you to move the laser head a few millimeters up or down. If you find a toolhead that allows you to adjust the focus of your beam, the laser can be mounted straight to the *Y-ax*is carriage. The laser should plug into your controller board in exactly the same way as the 3018.

In the next section, we will discuss endstops, much as we had with the 3018 machine. With a machine as simple as this, you may find these unnecessary because the machine is less likely to jam on the workpiece (it doesn't touch anything).

Endstops and electronics

A distinct item you have not seen added to this design is endstop switches. You can use exactly the same switches that you acquired for your 3018 and mount them on the extrusions as you see fit (do not use the switches that come on rod mounts; those switches are fine, but their mounts are not). I haven't done this here as they are not essential, and installation will depend on what you buy. I suggest you place them at the 0,0 position, and so the switches are at the left at the rear of the machine if you are looking down the *Y* axis from the *X-a*xis tensioner. If you look at *Figure 8.1*, this would be the corner of the machine at the bottom left.

For electronics, you will need the usual components – notably, the board, power supply, and optionally, an emergency stop switch and LCD controller. You can 3D print or fabricate a box to enclose the electronics and mount the box against the feet of the machine. If you enclose your machine, you can run the electronics to be housed outside the enclosure.

Just as with the 3018, you can flash the board with stock GRBL and then run exactly the same control commands and GCode-generating software for movement and laser control.

The same machine frame can be turned into a plotter with a few tweaks. Let's take a look at that next.

Plotter

The biggest reason I like this machine's architecture is its versatility. The same mechanical platform can be used for your plotter. There are plenty of designs on Thingiverse and YouTube that detail plotters of various capabilities that use the same cantilevered system we have here, but the biggest difference between them will be the toolhead. For the plotter, you can fabricate exactly the same machine or modify the same machine as previously to have a different mount for the toolhead. The plotter toolhead will need a small servo motor to raise the pen up and down as it draws on a sheet of paper. If you are feeling creative, you can create a mount for both toolheads. In preparing for this chapter, I actually made two machines and 3D printed the toolhead for the plotter from a design I found on Thingiverse at `https://www.thingiverse.com/thing:5149959`. I determined that you could easily hand-cut the same toolhead out of a piece of 3 mm plywood or MDF. If you do go down this route, I highly suggest you beef up the servo motor mount area and just make sure there is more material remaining after you cut out all the holes. The two gears this design requires can also be hand-cut, 3D printed, or cut with a laser. Interestingly, you can use your 3018 to cut the parts, or the laser cutter from this chapter!

The only other consideration is how you secure the pen or marker. This is going to depend on what you are using. If you stick with the Thingiverse design, you will either have to 3D print the pen/marker holder (and gears) or devise something suitable. However, these parts are so small and so simple, it is fairly inexpensive to get a set printed at your local maker space (likely for a nominal fee, or if you have a friend there, for free), or at any 3D printing service.

Here is the 3D-printed carriage on a piece of 2020 extrusion. I found it easy to drill an extra two holes in the middle of this and mount it on the *Y-a*xis carriage we already had from the laser cutter.

Figure 8.13 – 3D-printed plotter carriage

Summary

In this chapter, you have seen a design for a versatile type of purpose-built CNC machine that runs faster than a 3018 for laser cutting and plotting. The same machine also can be applied as a drag knife cutter. Because this is a belt-driven machine and the toolheads are very light, they can operate much faster than the 3018 for these functions. That is not to say that the 3018 in all those modes doesn't work, but you get added utility and the ability to run parallel, unrelated jobs rather than tying up your machine to run those same jobs in series. More importantly, this kind of machine can have its work area adjusted by simply changing the dimensions of the two larger extrusions. The limit to this is how much the machine can take the way it is without tipping over (the *Y-a*xis mount stabilizes it). However, I built a machine recently that adds a second 2040 extrusion, feet, and *X-a*xis carriage that supports the *Y* axis on both ends and there is no change in the software. You don't even need to add another motor (although you could).

Let us now move on to the next chapter, where we will look at creating a "poor man's 4th axis" for our 3018. This will allow our 3018 to cut things such as wheels, even with tread patterns!

9
Project: Building Your Own 4th Axis

Previously, we looked at a rotary axis for your 3018 that was commercially produced. However, we may want to create our own with its own particular features, including the ability to hold components much like you would on a lathe. Rather than depending on the symmetry of the object being machined so that it rotates evenly on rollers, you may want to work on parts that would otherwise not sit on a store-bought 4th axis in a stable fashion. An example might be a narrow cylinder shape, or one that doesn't have a consistent diameter (such as a bottle). To that end, we will look at building such a unit for your 3018 based on work by ZenziWerken. You can have a look at his original design here: `https://www.thingiverse.com/thing:2344975`. In this chapter, we will look at the design and update it as necessary for our purposes.

As part of this project, we will develop a few skills and recycle what we have learned with the store-bought 4th axis:

- Modify a design to suit the materials available to us and extend our knowledge of this very useful tool
- Learn how to use a different 4th axis
- Learn more about converting files for use in CAD

Design is the starting point

The design presented is made up of several components, mostly fashioned out of flat stock. Originally planned to be made out of 6.5mm plywood, your 4th axis structure will be fine with 6 mm (1/4") plywood from your local hardware store. If you decide to make yours using the 3018, you can use the original design, but you will have to find exactly the same components the original demands. However, in my case, I was unable to find the same parts and I had to improvise. I was also concerned that the gears in the original design would not hold up to extended wear. Because I needed to experiment, I converted the design to STL files so that I could 3D-print the parts as part of my prototyping effort. To that end, I had to modify the design by uploading the SVG file provided to TinkerCAD and make my modifications there. Also, because I needed to do some quick prototyping, I 3D-printed my parts. Let's start with what the original SVG file looks like when imported into TinkerCAD. When you load the file, you will see all the parts come in at the same time. Through experimentation, I determined that the ideal scaling level is 70%. So, when you do the import, tell TinkerCAD to bring the file in at 70% of the original.

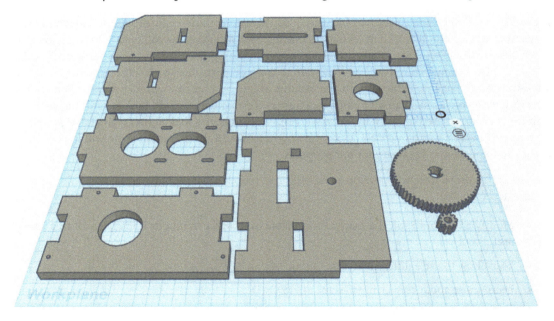

Figure 9.1 – The SVG file imported into TinkerCAD

The reason for scaling (for me, anyway) is related to how I pulled in the file and how TinkerCAD interprets it. Once you scale the file, you will need to carve out each part to fabricate it separately. You cannot fit all these parts on your 3018 work surface, so you will have to cut (or 3D print each part). If you decide to 3D print, I suggest you print the parts at 100% infill. I do warn that the fit, whether cut or printed, is very tight, and some filing of the tabs or slots is necessary. In my case, I used a heat gun to soften up the plastic a little (without deformation) to allow me to press-fit the parts together. That worked very well.

Create a separate STL for each part by inserting a transparent box around all the parts you want to remove. When you group the various boxes, you will have the parts separated.

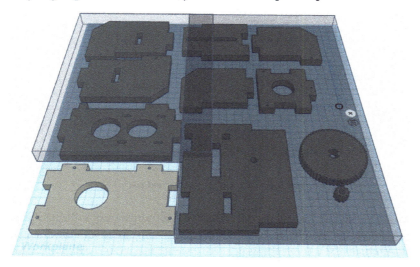

Figure 9.2 – Separating each part from the whole

When you group all the objects here, you will end up with a single part that you can process individually. This is important because, later on, we have to modify two of these parts to accommodate the availability of bearings I could find from online vendors. In the following diagram, you see the un-highlighted part above separated from the whole.

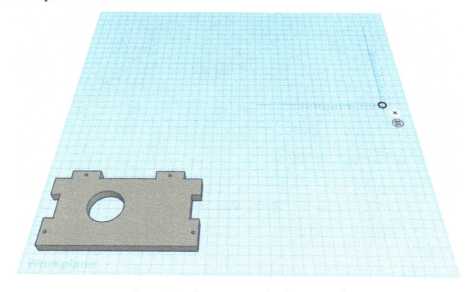

Figure 9.3 – One part completely separated

As you can see here, I now have this one part as a standalone component. You can repeat the process for all the parts by copying this one part, ungrouping the copy, and then moving around the highlighted parts:

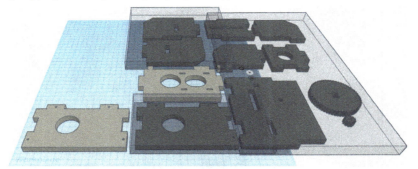

Figure 9.4 – Moving on to the next part

Go through each part until you have all the parts isolated. I refrained from utilizing the gears, which necessitated alterations to the smaller gear in order to make it compatible with the designated NEMA 17's motor shaft. Alternatively, if the gear was printed, it would need to be heated with a heat gun and pressed onto the motor shaft. Here's a screenshot of the second part I isolated in *Figure 9.4*:

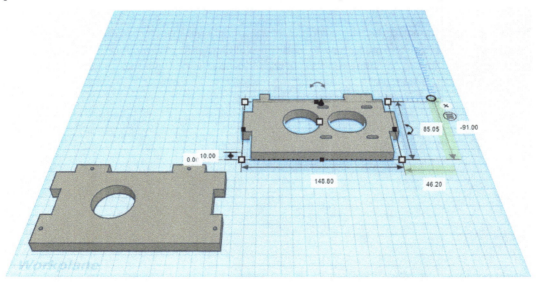

Figure 9.5 – The second part isolated

Important note

Note that you should NOT add the already isolated part(s) when you group so that you can move and manipulate each part separately.

With all the parts isolated, you can now proceed to modify the two parts that need to accommodate the bearings that I was able to acquire – notably, the same parts you see in *Figure 9.5*. In *Table 9.1*, you can see that I used two different types of bearings. This was mostly a matter of convenience because I had one of each available. You will only need to modify the two round (not the one oval) holes to accommodate your bearings. For the part you see in the preceding figure on the left, I needed to widen the hole to accommodate a 6000-ZZ bearing. You can see how I did this in TinkerCAD, by adding a cylinder of suitable size. The dimensions are obvious in *Figure 9.6*; however, note that the hole diameter needs to be 26 mm for this bearing to fit. To make that happen, as I was 3D-printing, I again used my heat gun to allow me to hammer the bearing in place.

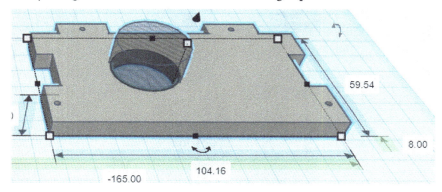

Figure 9.6 – Widening the hole for the 6000-ZZ bearing

You will need to do the same thing for the second part and make that accommodate a 608-ZZ bearing, as depicted in the following diagram:

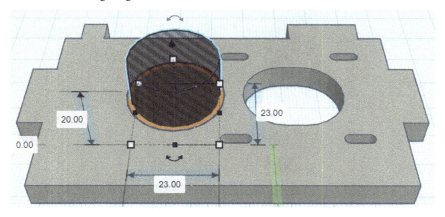

Figure 9.7 – Widening the hole for the 608-ZZ bearing

Note that, for this part, I had to experiment a little before I determined that the best diameter hole for my bearing was 23 mm. That was not a tight fit, but I secured the bearing with epoxy, which was very effective.

Having figured out all the details of the 4th axis design and made the needed updates, it is time to complete the axis and add it to the toolset for the 3018.

Fabrication

After several iterations, I was able to settle on a set of components for the project. Here is an image of all my attempts to get this right.

Figure 9.8 – Parts from the prototyping effort

Once you have all the parts fabricated, you can go ahead and do some test fitting before assembly. Here is my frame (I decided I don't need the "tail stock" portion of this for my purposes).

Figure 9.9 – Test-fitting the axis frame

You can see how I had to do a lot of filing to get everything to fit properly, but that's only because the tolerances are so tight. You can always experiment by changing the scaling to go down by fractions of a percent, but that seemed like a long and drawn-out process when I could get things done faster with some post-fabrication trimming.

Here is a parts list for the 4th axis. You should be able to use any NEMA 17 stepper motor, and you are likely to have a spare leftover from the project in *Chapter 8, Upgrading Your CNC Machine*.

Item description	Quantity	Source/link
6000-ZZ bearing	1	`https://www.amazon.com/dp/B07T59CWXX` You only need one bearing, but if you choose not to use this bearing, you will have to adjust the hole size on the front panel accordingly
608-ZZ bearing	1	`https://www.amazon.com/Bearings-Double-Shielded-Miniature-Skateboard/dp/B07H83VV6B/` You only need one, but these bearings are so useful for other projects
Drill chuck set	1	`https://www.amazon.com/gp/product/B09QSNZSQV` This chuck set is great. You will need the medium (8 mm) sized one of these largely because it has a longer drive shaft. Ideally, the larger chuck (10mm) would be better, but that shaft is too short. I was unable to find one in stock that the original designer had used.
Closed loop timing belt and pulleys	1	`https://www.amazon.com/gp/product/B07KBFWD6W` This timing belt set comes with two pulleys, the smaller 20-tooth pulley and the larger 60-tooth pulley. Buy the 8 mm version. That should allow the 60-tooth pulley to fit on the 8mm chuck shaft. However, the 20-tooth pulley will not fit on the 5mm shaft of the motor (as it is too big). This is why you need a separate 20-tooth pulley.
20-tooth GT2 pulley with a 5 mm bore	1	`https://www.amazon.com/WINSINN-Aluminum-Synchronous-Timing-Printer/dp/B077GNZK3J` You only need one of these pulleys.
8 mm collar	1	`https://www.amazon.com/Zeberoxyz-Stainless-Material-Isolation-Accessories/dp/B08SK2LNNV`
Set of closed-loop GT2 belts	1 set	`https://www.amazon.com/gp/product/B088M3V865`
M3 screws to mount the motors Suitable wood screws to secure the panels; if 3D-printing the panels, M3 screws will do fine. Alternatively, simply glue all the panels together.	Various	

Table 1.1 – The parts list for the 4th axis

Just as before I printed all the remaining panels and used my heat gun to press fit everything together. A dab of CA glue (also known as superglue) makes sure the panels stay in place.

> **Note**
>
> It is easier to mount the motor on its panel prior to assembly, since access to the M3 mounting screws will be constrained by the front panel (you can make that panel removable by only screwing it in place and not using any glue for it).

You can see what I mean in *Figure 9.10*. Note how little space there is to maneuver once the front panel is in place. The 8 mm chuck can be seen at the top-left of the photo.

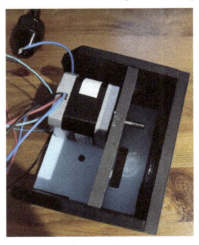

Figure 9.10 – Test-fitting the stepper motor

I am using an old stepper motor from a previous project. If you do the same, and the wires are not crimped into a suitable plug for your CNC machine's board, you will need to identify which two wires go together. That process is fairly simple:

1. Find a multimeter and set it to voltage detection.

2. Touch one probe to a wire and the other probe to any of the other three wires one by one. The wire that comes back with a "short" (i.e., a completed circuit) is that first wire's pair. These motors all have four wires and, therefore, two sets of pairs. Once identified, you can now crimp a plug.

Note that the motors you purchase from the parts list are all set with a plug, but just in case you take something out of a parts bin to create a plug, this mechanism should still work well. After I identified the wire pairs, I just stuck a piece of tape on the back of the motor to remind me. The motor direction can be reversed by simply flipping the plug in the CNC machine board's socket.

Figure 9.11 – Figuring out the motor wiring was easy

Next, let's discuss the installation of the 8 mm chuck. This chuck comes in two parts – the chuck head and the shaft that screws into it. When you screw the shaft in, you will need to take it only as far as you need to so that the 8 mm collar prevents the shaft from moving inside bearing. I would also advise you to apply some CA glue or epoxy to the shaft during assembly to lock it into the bearing. The collar is only a backup. Make sure there is a little space between the rear surface of the chuck and the front plate so that the chuck does not chafe the plate. Here's a photo from when I tested the chuck's placement:

Figure 9.12 – The chuck is placed in the 6000zz bearing ahead of securing it to the 60-tooth pulley

Here is a photo of the two 20-tooth pulleys you will encounter. Note the one with the larger bore. That's the one that came with the 60-tooth pulley.

Figure 9.13 – The 20-tooth pulley we will use is the one on the left. That has a 5 mm bore

And here is the 20-tooth and 60-tooth set I purchased. In the end, I had to widen the bore on the 60-tooth pulley to fit onto the chuck's shaft.

Figure 9.14 – The 20-tooth and 60-tooth pulleys and the closed-look belt

The belt that comes with the 60-tooth pulley is too long, but the set of closed-loop belts you acquire will be suitable (use whichever one fits your choice of pulley sets). I used the 158 mm one, which worked perfectly.

With all the parts test-fitted, it is time to finish our unit and put it into service.

The final assembly

We now can proceed to complete the assembly of the 4th axis (again, note, that I don't need the tail stock, but if you need it because you have a very long cylinder to machine, you should build it in exactly the same way we did the main component).

We start first by mounting the pulleys on the motor and the chuck shaft. We also need to install the 8 mm collar on the chuck's shaft, between the plates. Do not tighten the grub screws on the collar just yet. As an option, if you have enough shaft clearance, you can put a collar on either side of the plate to hold the chuck securely against the bearing, even though you have glued it into the bearing's bore. Here's how the 60-tooth pulley and collar fit on the chuck. This is just for illustration; you will need to assemble this element with the plates in place.

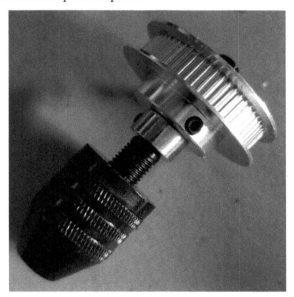

Figure 9.15 – The chuck and pulley assembly (without the front and rear plates)

Next, secure the pulley on the motor and slide the belt over it. You can now mount the chuck and the 60-tooth pulley, making sure the belt goes around the larger pulley. When in place, you will need to adjust the tension on the belt by adjusting the position of the motor. The mounting holes are oval for precisely this purpose.

At this point, you can go ahead and tighten the collar's grub screws. If you used a second collar to provide spacing for the chuck itself from the plate, go ahead and tighten the screws on that too.

For the rear of the chuck's shaft, I wrapped a small piece of tape, to enlarge the shaft's end just enough to be a tight fit in the 608-ZZ bearing's bore. A dab of CA glue or epoxy secured the shaft. From there, the only thing left to do is to attach all the plates permanently, and the 4th axis is finished.

Here is the fully assembled 4th axis. At this point, the only thing left to do is to establish the stepper ratios in the software.

Figure 9.16 – Assembling the 4th axis before the belt is tensioned

Our 4th axis unit is now complete, and it is time to put it into operation. For that, we are going to have to make a few minor changes to our 3018 machine and adjust some settings.

Installation and settings

Note that in the original post by ZenziWerken, his gear reduction is 6:1, while ours is 3:1 (60 teeth against 20 teeth). When you add in the circumference of your workpiece, you will be able to determine how much linear travel in the 3018's X axis each revolution produces. For example, using ZenziWerken's 40 mm wheel, we can use the following formula:

40 x π (or 3.1415) = 125.66 mm

Taken against a 3:1 gear ratio, it's 125.66 / 3 = 41.89 mm

Just as before, the motor plugs in place of the existing X-axis motor on the CNC machine, and just as before, the 3018 will think your round object is really flat and cut into it as such. Remember, if it seems like your axis is rotating backward, simply flip the plug on the 3018 board.

Whether you use your end mill or your laser, you can now fabricate on round surfaces that can be suspended above the work table.

Summary

We learned a number of things in this chapter. We, again, learned how to adapt a design for our own purposes and use some basic ideas to use parts we could find in place of parts we could not. We also had a sneak peek at what it takes to convert a CNC-ed part to a 3D-printed part. Finally, we assembled a 4th axis that we can now use to machine or engrave cylinders of various sizes. One thing we did not take into account in this chapter is nonlinear cylinders, and this project is really meant for surfaces that are consistent – that is, the cylinder's work area does not have a changing diameter. However, there is nothing stopping you from working on a cylindrical surface that does have a changing diameter – you just would have to set the CNC machine to work each linear segment separately.

Let us now move on to the next chapter and look at creating a "poor man's 4th axis" for our 3018. This will allow our 3018 to cut things such as wheels, even with tread patterns!

10
Project: Adding a Laser to the 3018

In earlier chapters, we discussed the use of a laser toolhead on our 3018. However, a laser toolhead adds weight to our X-carriage. A heavier carriage means our speeds slow down and adds complexity. Earlier versions of CNC microcontrollers had to have the laser plugged into a **Pulse-Width Modulation (PWM)** header on the board. There have been instances where hobbyists have attached a laser to the part cooling fan header of a 3D printer microcontroller, but for our machines, there should be a header specifically for the laser. Some boards have **Transistor-to-Transistor Logic (TTL)** headers/plugs that you can also use. One of my older machines has something like this. One of my laser cutters uses an older two-axis controller with a TTL port, which has two pins. A PWM port/plug has three pins. In any case, the microcontrollers we have been using have a PWM port for the laser already in place alongside the spindle motor header.

We have also discussed adding a laser to a 3D printer, even replacing the hot end of the printer with a laser. This is fairly simple to do with most consumer-grade, lower-end 3D printers. We will not cover this here, but it is worth noting that it is possible to resurrect an older machine and repurpose it.

In this chapter, we will continue to expand the utility of our 3018 machines and look at the mechanical aspects of installing a laser toolhead. We will cover the following:

- Look at different methods of installing the laser toolhead

- Have a brief overview of some powerful laser-cutting software you can run on your machine

- Have a look at various laser head installations on different machines that you can use as inspiration

Selecting a suitable laser toolhead

There are two basic traits I look for when hunting for a commercially available toolhead: optical power and required input power. Optical power tells me how strong the laser is and whether it can cut through the materials I am planning to use in as few passes as possible. The strongest toolhead I dare use at home is a unit with 15W of optical power. These days, you can purchase toolheads with 20W or more of optical power. These are mostly diode lasers that use one or more diodes to achieve the rated cutting power. When I first started out, a typical toolhead had a single diode for the beam. Today, you can purchase a toolhead that features a combiner that merges the beams of multiple diodes to increase power. This way, diodes that can generate 5W each can generate 15W or 20W combined for excellent cutting power.

Remember that we are talking about a laser head that we are adding to a machine here. There are machines you can buy online that are already capable of this amount of power and more (e.g., Glowforge, K40, etc.). Earlier, we built a cantilevered laser cutter using one of these toolheads and built its alter ego, the drawing machine/plotter that is essentially the same frame.

The optical power output for your toolhead should suit your desired objectives: what will you be cutting with your laser – paper, foam/depron, or thin plywood? Remember that higher-power toolheads will cut into thicker materials. Very high-power toolheads (e.g., 15W and above) can engrave some metals and even stones. Also, with higher-power lasers, you can set the desired power output frequently as a percentage of the maximum. 50% power on a 10W laser is 5W for example. This is why I try to get the *highest output toolhead* I can find for the price I can afford while still considering the weight and power consumption of the laser. Weigh all these factors before determining what unit works for you, and remember that the toolhead will likely run off a completely different power supply than the rest of your machine. For most of my own cutting purposes, I don't bother with toolheads with less than 7W of optical power, and if I am cutting wood that is 3 mm or thicker, I prefer the optical output to be higher than 10W. Remember, a lower-power laser will eventually cut through thicker material, but only after many passes, and for each pass, you will find yourself adjusting the laser height to get the narrowest beam. This is why I try to acquire higher-power units because you can reduce power from the maximum.

In addition to the all-in-one toolheads you can buy online from various vendors (e.g., AliExpress, Amazon, or eBay), you can buy toolheads that have a lot of capability and functionality from vendors such as Endurance Lasers (https://endurancelasers.com). I have two of their toolheads. One is an earlier 10W unit and another is a 15W unit I have yet to install on my machine (it is meant to be mounted on my Ox (a much bigger and sturdier CNC machine in my inventory). Endurance's units have a separate control box, switch indicators, and other features that far outstrip basic diode lasers. They are really meant for heavy-duty use, which is why I have them. Here is that 10W unit mounted on an Eleksmaker chassis with 3D-printed parts (the original hobby unit had a small 0.5W laser, which didn't suit my purposes).

Figure 10.1 – My older Endurance toolhead mounted on a hobby-grade chassis with 3D-printed parts

Looking at the laser toolhead, you can see that it is composed of multiple components that provide control of the laser. A system such as this will last a lot longer than most off-the-shelf lasers when heavily used. Here is a photo of the 15W unit from Endurance:

Figure 10.2 – 15W Endurance Laser toolhead and associated components

Note the control boxes and their complexity when compared to the unit we put on the cantilevered chassis earlier. Endurance also makes multi-diode units, but those are generally more expensive.

Since we are focusing on our 3018 machines, we want to select a lightweight unit (the one we used on our cutter is fine), but we also want to consider how we will mount it on our 3018. There are two ways to do this, and we will cover them in the next two subsections.

Mounting adjacent to the spindle

Remember that mounting the laser on the X-carriage and so adjacent to the spindle means that the carriage will be heavier and so will have to be slowed down to accommodate this. Here are some pros and cons of this approach:

- Pros:

  - Once installed and connected, it is a fixture of the machine. To operate, load your G-code and go.

  - Eliminates clutter: no connecting and disconnecting wiring and no need to shut down and restart the machine as you would when manually changing toolheads.

  - Installing different toolheads is possible if you have a generic mounting point. When you want something more powerful, match the hole pattern and install your unit with all its accessories.

  - Your existing air assist setup can probably be used for the laser just as easily without having to do more than repoint the nozzle.

- Cons:

  - Heavier means slower travel speeds

  - You may need to design a mount of some sort to secure the toolhead to the spindle

  - The area (and anyone nearby) would need to be protected from debris flying into the lens or diode assembly

  - You have to make sure your installation does not interfere with the operation of any endstop switches you may have mounted

Mounting in place of the spindle

Take a close look at how your spindle is mounted on your 3018. See the grooves in the mounting holes? Those are meant to accept a suitably sized rectangular/square toolhead – typically how your single-unit lasers come. You just tighten the clamp, plug in, start the machine up, and start cutting. Not all spindle clamps have these, but the clamp itself should be big enough to accept your laser toolhead. Make sure your laser will fit in the clamp. If your laser is too big, then you will have to mount it adjacent to the spindle. Here is one of my 3018s with the grooves in the clamp.

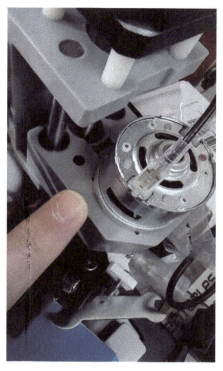

Figure 10.3 – Grooves in the spindle clamp. A square laser housing would slide in here nicely

To help with your decision-making, here are some pros and cons for this approach:

- Pros:

  - Lighter X-carriage

  - No need for additional mounting methods

  - Faster travel speeds

- Cons:

  - Accessories such as a dust brush will probably get in the way and will have to be moved/removed.

  - You are constrained to whatever fits in the spindle clamp.

  - The laser toolhead will have to be removed when not in use and stored somewhere (some users build a small storage mount/emplacement on the 3018's frame).

  - Depending on the spindle clamp, you may have to fabricate an adapter if you are using a smaller laser module. This adapter will have to fit inside the clamp's opening and the laser then fits into it.

For most lasers, you will likely need to have a separate power supply just for the toolhead alone. Just like we saw before, the toolhead will have the PWM cable, which goes to the controller, and a power cable or on-body connector for an external power supply. When you select a laser, consider making sure your power supply has an on-off switch as a safety feature.

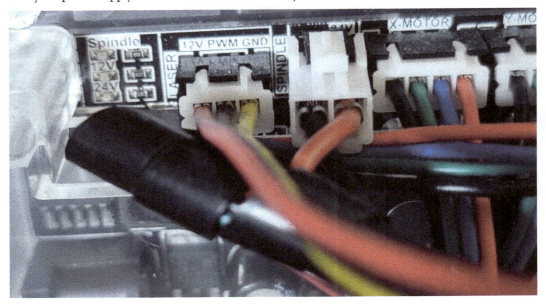

Figure 10.4 – PWM header on typical controllers. Note that both the
spindle and laser can be connected at the same time

Now that we have determined the dimensions needed for the laser module and how it will be mounted, we can shop for a unit online. There are several brands, some more well known than others. For example, I have two modules from NEJE, two from Endurance, and one from Ortur. Costs and laser types vary. Most of mine are blue light lasers with power output from 5W onward. I have paid anywhere from $50 to several hundred dollars (for the Endurance units). I always factor in the expected duty cycle of the laser (whether it is for hobby cutting or heavy-duty use, for example) before I decide on a module so that I can justify the expense. Sometimes, I will find a used unit on Craigslist, eBay, OfferUp, or similar sites just to harvest the toolhead from a cutter to use in my machines. Lightly used machines also provide parts other than the toolhead that are useful. There are folks who have built their own toolhead using Blu-ray laser diodes, but I prefer to buy something fully assembled so that I can get to cutting faster.

> **Note**
> Always be sure to purchase the appropriate eye protection to go with your particular laser. Blue light laser glasses are not interchangeable with red laser glasses, for example. Most laser toolheads come with a set of glasses specifically for that toolhead.

Let's now explore how you will control your laser cutter. Like everything else, there are several options with some staples. I am a very regular user of the first three of these. Their cost is reasonable and their features are extensive.

Laser-cutting software

Unlike standard CNC software, such as sending G-code using UGS, there are quite a few applications that are very specific to laser-cutting operations. I am a big fan of LightBurn, which is not free, but very affordable. There are others, of course, some free, others paid. Here is a list of some of them:

- **LightBurn**: My local makerspace makes heavy use of this software with a number of laser cutters, including some larger machines. You get to use it free of charge for a short trial period, but a license is very reasonably priced.

- **LaserGRBL**: This is free software, with some good features. This software is geared toward the hobbyist, making it ideal for many users.

- **T2Laser**: I have this installed on my computer as well as LightBurn because it has a great utility to flash the controller. It isn't free, but I have found it very useful with some of my smaller laser cutters.

- **BenBox**: This is also free, but I have found this to be temperamental depending on what microcontroller you are using. It also has limited features when compared to the others.

- **Inkscape**: This is not specialized laser-cutting software, but it does provide engraving capabilities and will run on a Mac. It is a free design tool that you can download and install.

Of course, there are other applications, but these are the ones I have used the most. You should experiment with various systems and see what works best with the hardware you have available.

In all cases, you would use the software to first connect to the laser, then check the border of the planned cut, and then start cutting. Unlike CNC, most times you can upload a JPG or an SVG file and have the software generate the toolpath.

We are now ready to mount our toolhead, and once again, we have to choose an approach that suits our purposes and our acceptable trade-offs.

Laser head installation examples

On one of my 3018 machines, I installed the laser toolhead on the side of the spindle clamp. The installation required me to drill a few holes into the clamp body. This allowed me to complete all connections on the machine and run it at reasonable speeds since the toolhead is very light. Here's a photo of how it looks:

Figure 10.5 – 3018 with laser toolhead installed on the spindle clamp

Notice the mount I am using here. It is a simple metal mount that is compatible with this particular toolhead. You mount that onto the clamp with screws and then a slider with a tightening screw allows you to hold the toolhead in place.

With this kind of installation, my laser is obviously offset to one side from the center of the spindle, so that means my starting point (the origin) is likewise offset. Keep this in mind when you use your workpiece because your machine's home position will not park the laser at the same origin point as the endmill in the spindle.

As you can see from the following photo, I am particularly fond of installing my laser right on the X-carriage:

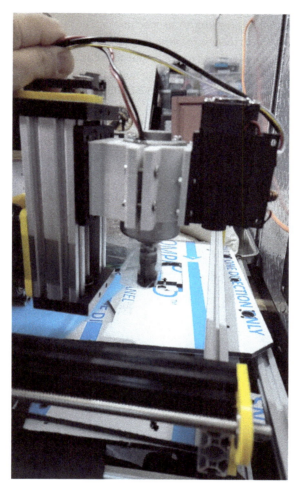

Figure 10.6 – BumbleBee's X carriage and laser head

The preceding photo is the setup on BumbleBee. This is a heavy-duty machine, with dual Y-axis motors and a very sturdy, all-metal Z gantry. Because of the structure's rigidity, it was an easy decision to mount the laser this way. The larger spindle motor and generally beefier carriage and clamp meant that I could have an installation that does not require me to plug and unplug components just to switch between the laser and CNC.

Summary

In this chapter, we have seen how we can turn our 3018 into a 2-in-1 laser cutter and CNC machine. Just as we added the ability to be a drag knife, our machine can be very versatile and cut a variety of materials using multiple methods. In the process of setting up the laser, we identified what laser toolhead we needed, determined a suitable mounting method, and added the components necessary to run the laser as an integral part of our 3018 machines. We also noted the software that we can use and looked at some example installations. Just as we can cut with a CNC on a fourth axis, we likewise can do the same with the laser. This means that our machine can really cut and engrave not only flat objects but also round ones. This means we now can engrave bottles, glasses, and anything else we desire. With a laser, we are not constrained to uniform surfaces since the laser will engrave or cut anything it can when it hits it. We have learned enough to allow us to build a larger machine, such as BumbleBee.

In the next chapter, we will discuss some of the particulars of a machine of this type and bring our journey full circle so that you can not only build but also customize your own machines to your purposes.

11
Building a More Capable CNC Machine

While we have focused our attention on the 3018 machine for most of this book so far, at some point, you will outgrow it. Maybe you need a bigger work area, improved metal machining capabilities, or greater accuracy. The 3018 is a very good all-purpose design, but by now, you will have seen its limitations, and if you have bigger projects in mind, the 3018 definitely will only be part of your overall toolchain.

When I first started with CNC, I built an airplane that required the fabrication of 52 wing ribs. Those were cut from sheet aluminum, and I had to cut the outlines and various-sized holes by hand. I hit upon the idea of cutting the ribs (and some other parts) with a CNC machine, and from there, I never looked back. If you are in the same boat, I am sure you are wondering, what next? So, in this chapter, let's discuss what you could do to machine bigger things. We have a couple of avenues:

- Improve and enlarge the 3018

- Build a purpose-built machine for heavier-duty work

- Achieve our goals through scale (i.e., more identical machines)

- Farm out the heavy-duty work to an external vendor and do the detailed work in-house

In this chapter, let's discuss each of these avenues and then explore one route I took – namely, building the machine I have referred to as "BumbleBee" elsewhere in this book. By the time we are done, you will have achieved the following:

- Examined alternatives with larger and more accurate machines

- Learned about the process I used to fabricate a CNC machine that fits a specific mission

- Determined the value of using outside services for more complex, or difficult tasks

Building a bigger 3018

Looking at the structure of the 3018, you can see that carriages are dependent on 8 mm rods (and, in some models, 10 mm rods). For short distances, this is fine. I have a Core XY 3D printer that uses 8 mm rods for the X and Y carriages that are some 24" long. When I upgraded the printer, I had to make sure that the X-carriage and Y-axis parts kept the assembly light enough to allow the rods to bend. Obviously, the longer the rods, the more there is a propensity to bend – and even a little bending is not good. The 3018 adds rigidity by having two rods, but even then, there is a chance, under load, of the rods flexing just enough to make your cut inaccurate enough to matter. That more likely applies to really long lengths of rod, so for the 3018, if you want to use the same method for motion, I suggest you do not make the work area more than twice the base size.

There is another factor to apply here. Even if you can use rods that are stiffened by a support rail (such as those you can find on Grainger's website – search for `Linear Support Rail`), you are still carrying a lot of weight on the Y axis with a larger table, a larger number of bearings, and of course, a bigger leadscrew and leadscrew nut. As your machine gets larger, it will demand a larger motor (such as a NEMA 23 versus the NEMA 17 that most 3018s have now), or twin Y motors to help move a heavier carriage. Bigger motors mean bigger current draws, which may, or may not, be supported by your original controller. Of course, your machine will also get heavier with the longer structural components and the supports for them. You will also have to work in more support to address any twist in the chassis, and if your 3018 uses acrylic gantry parts, you should replace those with their metal equivalents.

All in all, your 3018 may get bigger, but there is clearly a limit to how much you can "grow" it using the current motion system:

- **Linear rails**: Linear rails are great. They are strong, heavy, very accurate, and less likely to bend under load. Anytime I am able to, I convert machines with rods to machines with linear rails. This gives me rigidity and accuracy, but then the machine gets very heavy, and just about all the motion parts will have to be revisited (i.e., fabricated) if you are not buying off the shelf. Here's an example rail from my supplies. The little rubber inserts you can see in *Figure 11.1* are meant to stop the carriage from falling out. I have a lot of them on this rail because I harvested them from rails I had already installed. On a linear rail, you do not want the carriage to come off the rail because the little ball bearings in the carriage can (and frequently do) fall out and are hard to put back in.

Figure 11.1 – An MGN 12H linear rail meant to go on one of my 3D printers

- **Larger motors**: These will add to the redesign requirements. The leadscrews will have to be repositioned as the engine mounts shift to something larger, and of course, a single leadscrew may not be enough (leadscrews can also bend). If you don't want to use larger motors, maybe the answer is to use more than one motor for a given axis (typically, the Y axis), such as two NEMA 17s and two leadscrews, instead of one large NEMA 23 motor and a much larger leadscrew.

- **Ballscrews**: These are the 3018 leadscrew's big brother. They are much heavier, more accurate, and will further displace your original components and require additional design work. I'll give you a ballscrew example. Ball bearings flow inside the ballscrew nut, but you must be extra careful to not let the nut come off the leadscrew; otherwise, the ball bearings will fall out, and putting them back is very difficult. Have a look at this example, and note how much larger the leadscrew nut is as well as the leadscrew itself. I have placed it next to a TR8 leadscrew, similar to the one most of us have in our 3018 machines.

Figure 11.2 – A ballscrew meant for my much beefier Ox CNC
machine. The standard TR8 leadscrew is on top

- **A more rigid Z axis**: The plastic clamp would have to be revisited. Maybe, if you use rods, you want to try the metal clamp and carriage assembly. Alternatively, as you will see in Bumblebee, you can have an entire carriage assembly made of aluminum extrusions.

- **A bigger and more capable spindle**: Adding a larger spinning mass to your X carriage means a more solid assembly, since the spindle's weight could cause weaker/softer parts to flex and tilt your spindle ever so slightly forward, completely wrecking the accuracy of your machine.

- **V-Slot rails and wheels instead of rods or rails**: Just as you will see in BumbleBee, the chassis itself provides the rails. Not as accurate as linear rails but the tradeoff between rigidity and weight was worth it to me.

All these goodies, if we were to implement them, would completely change the 3018, and rather than trying to upgrade every little aspect, it's worth the effort to do a clean sheet design. Then, you will have a machine that is designed for a specific purpose, instead of trying to retrofit a hobby-grade machine. In particular, the gantry structure will take away from the rigidity of the machine and introduce flex to the chassis – the enemy of accuracy. That suggests we really should consider a compromise….

In the following subsections, we are going to understand how you can upgrade and enhance your 3018.

Standardizing and simplifying your structure

When you look at the 3018, you may see multiple extrusion types, gantry ends, the rod holder, and non-integral motor mounts. All this adds up to a lot of pieces and a high part count. One way to simplify your chassis is to reduce the part count, which is why I like the BumbleBee design. When you look at the design, it uses the same extrusions throughout (20x40 V-Slot), and those extrusions themselves are the rails. No rods, no rod mounts/holders, and less complex assemblies. If I wanted to build 10 of these machines at the same time, it would be fairly simple because the extrusions can be purchased pre-cut; the few yellow parts you see printed (or machined) and everything else is largely off the shelf.

Here's an overhead view of BumbleBee while it was being assembled. The design is very simple, and all the extrusions are the same length.

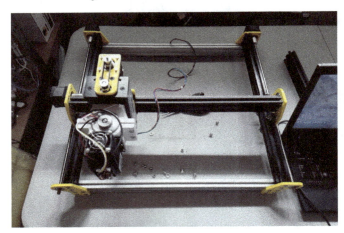

Figure 11.3 – An overhead view of BumbleBee. Note that all the extrusions
for the X and Y parts of the chassis are exactly the same length

Sure, wheels wear down faster than rails, but for the amount of effort to build the benefits outweigh the compromises. Using leadscrews throughout means the machine will have a slower head travel, but with a large spindle, that might be acceptable. Even with the laser head installed, we can dial down the power to accommodate the slower travel speeds you get from leadscrews (especially on the X axis). All this is fine because we can expect to machine harder materials with this machine than what the 3018 might have to work with.

Eliminating the moving table

One way to reduce the weight of a machine is to make the X carriage move along the body of the chassis. This means that rather than having a table move in Y, as we have with the 3018, we instead move the entire X carriage in Y. In the 3018, the workpiece moves in the Y axis, but for this approach, the part is entirely stationary, but the toolhead moves around it. This is what I chose for BumbleBee. If you look at *Figure 7.2* in *Chapter 7, Enclosures*, you will see an example of this in my 10-W laser (I mentioned it previously in other chapters as well). That laser's entire X carriage moves along the workpiece. In other words, the workpiece is always stationary, but the X and Y carriages move back and forth along it.

Another benefit to this sort of structure is that I can clamp the workpiece and any waste material to a table (even a really large piece) and move the machine around its surface – I don't have to cut up the workpiece to something that fits on a moving table like the 3018.

On a structure like this, you can use the rigidity of the chassis to prevent any deflection in your rail. You can use whatever rail system suits you fancy, as long as it is directly bolted onto the chassis members. Linear rail supports, linear rails, or V-Slot wheels (which require the chassis to be made of V-Slot extrusions) all work fine. BumbleBee uses V-Slot extrusions so that the carriage rides on wheels that roll right on the chassis itself – the extrusions themselves are the rail – which is simple, rigid, and very effective. In fact, BumbleBee's big brother, the Ox, with the half-horsepower router as the spindle, uses extrusions like this. Note that the more wheels you have, the better, but also note that you need to use eccentric nuts to make sure there is no "wiggle" in the wheels, and that they are firmly in contact with the chassis extrusions on at least two sides. The following figure shows the wheeled carriage system for both the X and Y carriages:

Figure 11.4 – BumbleBee's V-Slot chassis with their X and Y carriage wheels

Note how the chassis itself is the rail, making for a robust and lightweight motion system.

Use multiple motors

When I was building BumbleBee, I set a rule for myself to have a design that required me to purchase as little as possible and maximize the contents of my spare parts bin. BumbleBee's original design is on Thingiverse at `https://www.thingiverse.com/thing:3833005`. The original design had plates made of wood, and later metal, but I started with yellow PLA for the parts at 100% infill (i.e., fully solid) and added black extrusions (which gives me the black and yellow color of a bumble bee, hence the name). I had to revisit some of the plates to make everything work, but the intent was to have the machine make its own replacement parts out of aluminum, replacing anything yellow with the metal equivalent – one part at a time. One of the reasons I like this design is that it allowed me to recycle a pair of NEMA 17 motors for the Y axis, along with a pair of leadscrews from my spares. This design uses Delrin leadscrew nuts, but I could have easily derived a method to secure the typical brass nuts to the various plates instead (so yes, I did have to purchase the Delrin nuts, but on the plus side, they are self-lubricating). Note, in the following picture, the flexible coupling on BumbleBee's Y motor. This allows any alignment or flexure in the leadscrew to be absorbed by the coupling and function as required.

Figure 11.5 – A close-up of one of the BumbleBee Y-motor and leadscrew
assemblies. There is one on each side of the chassis

An all-metal Z axis

Most 3018s feature a fairly simple Z-axis assembly that does the job very reliably and is frequently made of the same heavy-duty plastic the clamp is made of. Of course, some heavier, more solid carriages are all metal, but most of the time, I encounter 3018s with a lot of plastic parts in Z.

However, for my BumbleBee, I wanted a much more rigid assembly that did not rely on rods and could be modified to be larger should I ever choose to do so. I also wanted something that, when mounted, was unlikely to cause any twist in the X-axis, which is why I assembled BumbleBee's extrusion-based Z. Granted that while I did have to purchase some parts, most components (such as the extrusions and wheels) are commonly available parts. The design also still allowed me to use the same drive system for Z that the original designer had. I did, however, dispense with the carriage as designed and developed a method to mount my much beefier Z assembly on the same carriage. In the following picture, you can see the Z carriage as well as the much larger combination of spindle clamp and laser holder.

Figure 11.6 – A close look at BumbleBee's Z-axis motion mechanism. We saw
this photo in the previous chapter, but it bears repeating here

As you can see, only the clamp is plastic, but I spared no amount of plastic to ensure it is as solid as it can be when I 3D-printed it.

A more robust controller board

BumbleBee will have an MKS DLC V2.0 board (there are later versions of course, but recall my rule of purchasing as little as possible), and an MKS Base TFT 3.5 touchscreen for offline control. This arrangement allows for the same features as what you have on your 3018, but the board does allow you to install your choice of motor drivers for smoother and silent operation, and the touchscreen has a very intuitive interface with an SD card slot.

There are cheaper boards from a variety of vendors, and you will find no shortage of these online. My rule for low-cost boards is that they should still be supported by the vendor, support the X, Y, and Z axes, and support at least a spindle and laser setup. I try to avoid spending more than $50 for most machines, such as a 3018. SainSmart and Genmitsu have a number of options you can choose from around that price.

I 3D-printed for the controller board, but you can just as easily use the 3018 itself to cut some flat stock material that you then assemble into an enclosure. I don't have a design, but with some basic design work, you can modify enclosures found on any of the maker websites such as Thingiverse, Printables, or Cults 3D. Here's what my lash-up looks like. I probably should have painted the enclosure black or yellow, but I ran out of yellow filament when I printed it.

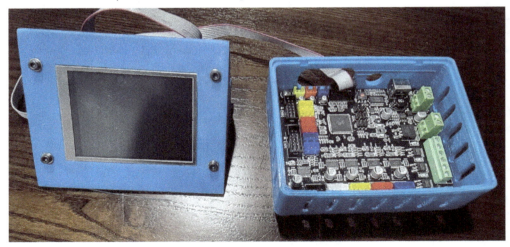

Figure 11.7 – The MKS DLC controller in its enclosure

These controllers are incredibly inexpensive – $40 can buy you the controller, touchscreen, driver, and cables on AliExpress. Some boards even have Wi-Fi connectivity so that you can remotely control them. Note that my version does not, but the touchscreen will accept a Wi-Fi module to allow for remote control.

Scaling up

I very much tend to geek out when it comes to machines. Sometimes, I build a machine and solve all its problems just to learn what works and what doesn't. At times, I will rebuild a machine many times over, with each generation better than the next, even though the underlying design is largely unchanged. This is why you will have seen multiple 3018s throughout this book. My first was one that I 3D-printed parts for, and I made most of it out of scrap materials. From there, I built up others, and now I have four of them.

If I wanted to fabricate many examples of a specific part (for example, when I need eight identical brackets), I can either try to fit as many as I can on a workpiece or print a smaller number but on multiple machines concurrently. By splitting the work across multiple machines, I cut the fabrication time drastically. If it takes two hours to cut one part and I need, say, four of them, I would cut all four parts in two hours instead of eight. If I pressed all my machines into service, I could scale up fabrication to make as many parts as I ed for a given project in a parallel (versus serial) fashion. This way, I don't have to wait for my CNC machine to have all the parts I need for an assembly and can get on with building the thing I am working on sooner. Obviously, I would need to make sure that all the parts are cut with the same relative accuracy across all machines.

If you intend to make a business out of your efforts, you definitely will want multiple machines for multiple purposes. As a general rule, you should build in redundancy in your shop wherever possible. If you will be cutting for profit, make sure your 3018 is backed up by another or something bigger so that an inoperative machine does not hold up your productivity. If you intend to make parts for clients in bulk, then having multiple machines working simultaneously ensures you can complete projects faster (and get paid faster), even if a machine goes down for whatever reason. You should expect machines to routinely go down for various reasons when heavily used (e.g., burned-out spindle motors, broken end mills, or controller board malfunctions). This is also why I keep a supply of spare parts for each machine readily available. Just like my 3D printers, the first thing any machine fabricates is its own upgrades or spare parts, all of which go into a bin, ready for the machine to be put into service as soon as there is a breakdown.

As hobbyists, there is only so much scaling up we can do. When it becomes untenable to fabricate certain parts (such as when machines cannot handle materials efficiently, or the desired workpiece is too big for our shop setups), this is when we will have to go to an external vendor. I have mentioned before that some companies will cut all sorts of materials to your specifications using your original digitized models fairly efficiently. A lot of them use industrial waterjet cutters and can run through jobs very quickly. The enclosure parts for one of my lasers are too big for what I had in my shop at the time and were cut by such a vendor. I didn't even have to order the material separately, as they supplied the raw material as well. I only sent them the digital models (DXF files in this case), and a couple of weeks later, a package arrived with all the parts cut perfectly. Here is that enclosure post-assembly, with the laser electronics sitting on top.

Figure 11.8 – The laser enclosure. Note that the polycarbonate
window has not had the protective film removed

This enclosure started life as intended for a 3D printer and was to be made of plywood, but a few minor modifications to the assembly made it a very viable laser cutter enclosure made out of 2-mm aluminum sheet. The hinges and knob on the lid are 3D-printed, but I could just as easily have purchased them from my local DIY store.

Summary

By now, you should be able to upgrade and enlarge your 3018, or even have it fabricate parts for a complete replacement of itself. There are a lot of designs on various sites (Thingiverse, Printables, and Cults 3D) for various CNC machines. You can pick one that suits your application and budget or design your own. Unlike the 3018, you now are not confined to a single motion system or even an electronics suite. Once you have established your requirements, you can move ahead with your build. As you have seen here, it is also possible to have a machine make its own upgrade parts in a steady path toward continuous upgrades.

You also have learned the potential of scaling up your capabilities to run multiple machines concurrently and establish redundancy in your shop. We also touched upon the benefits of sending off some fabrication to a third party, just to make larger parts before you refine and rework them further.

As we approach the end of this journey together, we can now bring everything together in the final chapter of this book and discuss some of the more sophisticated machines that you will find in the industry, which, if you are feeling adventurous, you can explore yourself – for example, using a fifth axis in CNC, or having an electronically controlled lathe that carves a cylindrical workpiece through G-Code

12
Future Projects and Going Bigger and Better

It has been quite an adventure so far. We have gone from a simple bare-bones desktop CNC machine to machines that can cut, draw, and carve with end mills, lasers, pens, and knife points. We have added all sorts of upgrades to our desktop machines and assembled bigger, purpose-built machines.

For our final chapter, we will go over some ideas for even larger and more complex machines that require some additional expertise, electronics, and specialized software. We will start with a scaled-up CNC machine and move on to some more fascinating ideas, such as the following:

- The ShapeOKO line of CNC machines you can purchase that go beyond the basic desktop hobbyist market

- A brief overview of my upcoming **OX** build and what I have planned for it

- Another brief overview of hot wire cutters, typically for cutting foam

- A look at automating lathe work

- How a fifth axis takes advantage of additional motors so that a CNC machine can address every surface of your workpiece (so far, we have only explored the surfaces above the work area – that is, the table on which the material sits – which means it has to have at least one flat surface)

- **Selective Compliance Articulated Robot Arm** (**SCARA**)-type machines that hold the workpiece in place but allow the toolhead to move *around* it to perform its machining tasks

As someone fascinated with mechanics and with a passion for building things (both kinetic and static), I find that CNC is only part of the equation in today's world. Nowadays, we have an array of automated, semi-automated, and completely manual tools and methods to fabricate objects, using just about any material. CNC is a subtractive (meaning it removes material, much like a sculptor) technology, and we also know about additive technologies, such as 3D printing, which can be done with plastics, pastes, and even organic materials. There are a few more cutting technologies we didn't cover here in CNC (such as hot wire foam cutters and lathes), and we didn't discuss forming machines that use vacuum and heat to form plastic parts, or machines that stamp-form metal. Those are not so much within the realm of CNC, but I mention them here because they are tools you should consider for your shop so that you can create objects at scale and in one step.

By the end of this concluding chapter, you should have all you need to equip your shop with all the CNC machines and their derivatives:

- You should know what your shop needs, and whether you need to make or buy it
- You should be able to track and participate in trends with CNC machining even if you don't have the capability in-house
- You can learn more about going beyond hobby-grade machines
- You can look at automating machines available in any shop (such as a lathe)

The ShapeOKO

Long ago, I purchased a kit for the ShapeOKO 1. At the time of this writing, they are on version 5 of this machine. This unit is more a "pro" unit that goes beyond the desktop hobbyist but is still small enough to occupy a suitable spot on your desk (albeit, more space than the 3018). The ShapeOKO 1 used belts for movement, using all-metal components. I outgrew mine almost immediately, and it has sat languishing and waiting for its upgrades. The new versions of the ShapeOKO have features that improve rigidity and accuracy that we have discussed elsewhere in this book, such as rails and ballscrews. I will eventually modify mine to use leadscrews and rails and replace the current contrller with an updated controller, or convert it to a purely laser platform. The ShapeOKO is expensive when compared to our 3018s, and the jump in cost is significant. However, it is a more "serious" machine, and if you are doing more than hobby work, this machine or one of its competitors is definitely worth a look. Here's a picture of my poor ShapeOKO 1, still waiting for its upgrades.

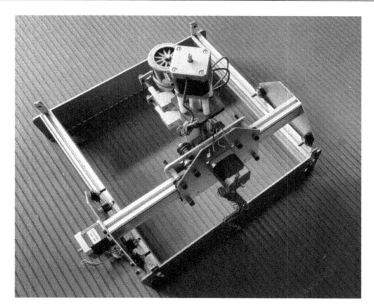

Figure 12.1 – My old ShapeOKO 1 waiting for its upgrades

If you find one of these machines somewhere online or you can replicate the design, do consider it as a mid-range unit or a stepping stone to even larger machines.

The OX

I first encountered the OX CNC machine at my makerspace when one of the makers was offering pre-cut metal kits for it. I jumped on the opportunity after I had a chance to do a little research. The OX is an open source, heavier-duty machine that you can find on the OpenBuilds site at `https://openbuilds.com/builds/ox-metal-cnc-router-mill.3681/`. The Thingiverse site for this machine is at `https://www.thingiverse.com/thing:1660320`.

As machines go, this one cries out for ballscrews (or heavier-duty leadscrews) very loudly. It is large, heavy, and meant for serious work, such as milling harder materials. Mine has a half-horsepower table router motor as the spindle from DeWalt, but if you decide to build one of these, you can use whatever takes your fancy. If your 3018 is up for the job, use it to cut the carriage pieces for you, or grab the drawings and either hand-make them or send them out for fabrication. There are some parts that need to be 3D-printed, and the brackets holding the extrusions together can be purchased online (aluminum or black anodized) inexpensively. When you look at the site, you will see some attributes that set this unit aside from the smaller machines we have seen thus far:

- **Larger motors** mean more power/torque and the ability to move a heavier carriage and toolhead. These are NEMA 23 motors. Four of them will make this machine substantially hefty and are meant for the workbench rather than the desktop.

- **Bigger, longer leadscrews** mean more travel, but also a propensity to flex, especially with the 8 mm ones when you get beyond 1,000 mm in length. Keep that in mind, and consider revisiting the design to use ballscrews or at least larger 10 mm leadscrews. Your 3018 likely uses TR8 trapezoidal leadscrews. For an OX, I would suggest you consider using TR10s. Those at least offer more strength, but they do add weight. On the Thingiverse site, the files offered include the option of using 4 mm or 6 mm thick material. I would suggest you stick with 4 mm initially and have the machine itself fabricate the 6 mm versions if you find it necessary. If you dig a little into the comments section on the OX site, you will see some users who have used 1605 ballscrews. These are heftier still, but you do have to put in a little design work to modify the leadscrew mounts to compensate for their shape and size. If you can come up with a design that assembles out of flat stock, you might even be able to fabricate most parts on your 3018.

- **More wheels** mean more rigidity, but they also mean specialized V-Slot-type extrusions. We saw this on my BumbleBee. Using rails here might pose a problem, since you may find it cost-prohibitive to have rails that are as long as the OX demands. Also, should you decide to expand your OX in the future, you would have to replace the rails. I do suggest you make sure your wheels are always in good shape before any run and that you keep several spares if you need to replace them. This is one machine I would not use rails for, just because of flexibility and cost.

For the 3D-printed parts, I would recommend you use a **polyethylene terephthalate glycol** (PETG) filament. If you feel adventurous or want something more exotic, there are metal-infused or even carbon fiber-infused filaments that would make for some very strong parts. My 3D-printed parts are made of PETG, but recently, I have been working with more exotic filaments that suggest I revisit some pieces.

One thing I like about the OX is that it is easily at home using Universal G-code Sender to control it. A machine this large merits its own PC. I like to use smaller footprint machines that I can mount under or to the side of the machine (these are typically all solid-state computers but are fully fledged desktops). My 10W laser, for example, is controlled by an Intel Compute Stick computer running Windows. The whole arrangement sits on top of the laser enclosure with its own Bluetooth keyboard and integrated touchpad. The monitor is an old laptop LCD encased in a 3D-printed frame, and the Compute Stick is mounted on the back of the LCD. Because it is a PC that controls the laser, I can store all the G-code files I need on my network and easily access them, without having to put anything on any other storage device.

Hot wire foam cutters

I've always liked building RC model airplanes. Many designs are available for free online, and from those, I've always wanted to build bigger models made from XPS foam. We discussed using our 3018 machines to mill blocks of foam, but if all we need to do is cut complex shapes without the mess that comes from the spindle, the way to do this is with a hot wire cutter. This is nothing more than a length of wire between two contacts that has sufficient current flowing through it to warm the wire. The heat from the wire melts the foam as you cut through it and allows you to shape it, with almost no mess at all. I have a hand cutter that operates as a wand, but I've always wanted to build something with automation. Here's what my handheld unit looks like. You can also make a handheld cutter using basic materials and a suitable power supply. Just look for instructions on the internet; there are many articles out there.

Figure 12.2 – My handheld foam cutter – not quite a wire but it works the same way

As you can see, a steady hand and clearly drawn lines on my foam part are key factors to ensuring I have a consistent cut through the material, needing little cleanup. Conversely, something that could cut precisely with a longer "wire" would allow me to cut through larger and thicker pieces of foam. This means I could shape larger parts instead of creating a skeleton that I need to later cover with some material. Foam can also be covered with fiberglass to add strength and durability. This is how the homebuilt airplane I am working on is designed. The wings are shaped with ribs made of foam (thereby creating a foam core), and the wing is then covered with a wet layup of fiberglass to create a very light and strong structure.

As always, I go to my stalwart website on Thingiverse and search for `hot wire foam cutters`, and I can see a variety of them shown in the results. Some are just handheld units, while a few are automated. There are some interesting ones, like these:

- `https://www.thingiverse.com/thing:5323256`, with a good video on its operation at `https://www.youtube.com/watch?v=ShMq1REG4eE&t=328s`

- `https://www.thingiverse.com/thing:2834138`, with a video at `https://www.youtube.com/watch?v=Oe-pj3NSOOw`

There are other hot wire foam cutters, of course, on Printables and Instructables, but these two illustrate the basic concept – two towers move together while raising a wire fixed on a gantry that moves up and down on each tower. As the towers traverse their rails, they drag the hot wire through the foam and cut the shape. The unique aspect of these machines is that they don't just go one way across a workpiece end to end – they can make one cut and then cut underneath the first cut, as if carving the airfoil for a wing. As befits a site like Thingiverse, most parts are 3D-printed, but you can always have that done for you. With a little design work, you should be able to fabricate a lot of the parts using your existing 3018 (the tower and rail gantries, for example). With a little careful study, you may be able to use off-the-shelf parts, as we did for our laser cutter for our moving platforms. For example, the same carriages we have on our cutter may be adaptable for the Y and Z axes. In effect, all you are doing is rotating a part of the laser cutter 90 degrees upward and placing two carriages on the Y axis instead of just one. The first design, with its gravity-based tensioning system, may be trickier to replicate using ready-made parts, but maybe you can use a spring-based tensioning mechanism instead.

CNC lathe

I've always wanted to add a lathe to my shop. I have a mini-lathe that I use for machining light woods and plastics, but nothing that can handle larger, harder materials, and certainly nothing automated.

I recently came across a maker on YouTube called Melkano. While the videos Melkano posts are in French, each system he develops has a site in English and French that you can follow. The designs he has are very easy to understand. I have already built one of his early mini-laser cutters, using some scrap extrusions and a few miscellaneous parts. Melkano also has designs for a CNC machine (very similar to a 3018) and a 3D printer. However, what caught my eye was an automated lathe. Unlike all our other machines, the lathe has the workpiece turning at a high **revolutions per minute** (**RPM**) rate, while a sharp edge moves along one side and cuts into it. Just think about how the vertical supports of stair rails might be made.

Here is a video of Melkano's lathe: `https://www.youtube.com/watch?v=4p3GYF2-vvE`.

What I love about this thing is that, unlike others that you can make, this one really can machine metal. I frequently have a need to machine cylindrical objects and have had to resort to 3D-printing them, which is not ideal for every use case. However, with a lathe like this, using the same sort of electronics in our laser cutter and 3018 machine, I can automate the machining of cylinders, while still retaining the ability to manually control the lathe.

Have a look at the companion Thingiverse site (`https://www.thingiverse.com/thing:5862193`) where Melkano has posted his parts. If you are unable to print the parts, Melkano also offers the parts for sale in his Etsy shop. Also on the Thingiverse site are files that you can either send to a PCB fabrication facility or use to make your own PCBs, using some additional components that are mounted/connected to an MKS DLC 2.0 board, which you have seen me use before (this is the primary controller for the lathe). The video is fairly easy to follow, and the machine looks very solid. This lathe is definitely in my build queue.

One idea I have is to convert some of the lathe plastic parts to aluminum using some simple casting techniques. I am particularly interested in the larger housing parts that may be cast via the lost plastic method. Since I only need one unit, I can use plaster or sand casting to make the housing from the 3D-printed parts as the mold master. I might also do the tailstock housing the same way (this part is the one holding the drill chuck at the other end). Alternatively, I could stiffen the larger 3D-printed parts by painting them in epoxy, which would further harden them. My ultimate objective would be to make this lathe strong enough to machine soft metals, wood, and also plastics.

A fifth axis

Machines with five axes are unique in that they are able to carve/cut not just on the upward-facing surface but also on all surfaces, bar the one where the workpiece is held. If you think about the engine block of your car, that likely was CNC-machined this way. There are some great videos out there of this, such as the following:

- This video takes a huge aluminum cylinder and turns it into a miniature V8 engine block. Look at all the coolant it uses, and as it carves and cuts, the machine also switches toolheads for each specific job. On top of this, not only does the machine drill holes but it also taps them – spectacular! Here's the link to the video: `https://www.youtube.com/watch?v=wHstzxuryMk`.

- Remember our chapter on the 4th axis? Here's a video that cuts an engine block with four axes, but unlike our modification, you can't sacrifice one of the axes for it. Here, you have X, Y, Z, AND Theta (the degree of rotation). No, it's not a five-axis machine, but it gets closer: `https://www.youtube.com/watch?v=USIMawXJsKQ`.

- This is a crazy video where the 4th and 5th axes are both rotational but on a gimbal. The purpose is to be able to carve down and sideways to create channels in the engine block. This is a full-size engine they are making here, and the idea is to be able to rotate the workpiece up, down, and around its lateral axis. For CNC nerds like me, it's really fun to watch: `https://www.youtube.com/watch?v=PWJAeEzqdc4`.

There is nothing stopping you from making your own machine that does something like these machines, but keep in mind that you are going well past the hobbyist stage here. There are CNC-ers who have made desktop-level machines that add the extra two axes with, basically, a turntable (instead of a stationary work area) that itself rotates around some axis. I did some digging around and found a couple that interest me:

- Here's a summarized video of a machine that I have been looking at for a while. It's a desktop unit and uses mostly off-the-shelf parts. Granted, you are going to have to fabricate some components, but it is an interesting study: `https://www.youtube.com/watch?v=IYl_cUHEWrk`

- This video is interesting because it is a DIY implementation of a five-axis unit that is largely 3D-printed. What makes it unique is that it uses electrochemical machining. It's like reverse electroplating because you remove material through the same process. While limited to conductive materials, you can machine very hard metals this way that may be difficult to machine through mechanical means. This machine uses straight Marlin (largely what most consumer-grade 3D printers use) as firmware. The video has links to parts that can be made and the firmware mods that are needed: `https://www.youtube.com/watch?v=KieJN-J4s38`.

- Another interesting machine is this one from Instructables: `https://www.instructables.com/DIY-Desktop-5-axis-CNC-Mill/`. What is interesting about this one is that it turns your 3018 into a five-axis mill with a few more parts. However, you still must revisit the controller board and the firmware, but the article is clear on what you would need to acquire. It doesn't give you a big work area but does make it possible to mill more sides of your workpiece.

- The machine in the next video is something I have been dreaming about for some time. The extra two axes are built into the Z gantry because, rather than the workpiece rotating (which stays stationary), the toolhead rotates around the material being machined. In this video, the CNC-er has a room-sized unit that essentially carves out an entire boat for him. The reason I like this is that I have been looking at building my own replica car (I happen to like the Lamborghini Countach, but can't imagine buying an original when I can try to make one). My goal would be to machine the entire body shell in 1, 2, 3, or 4 segments out of foam, and then lay it up with fiberglass, adding a metal skeleton. I could also machine the entire shell and turn that into a master mold for a carbon fiber/fiberglass mold, allowing me to make my own body shells and "renew" my DIY car as often as I would like. Take a look: `https://www.youtube.com/watch?v=nITLI_WcnuM`.

Of course, if you have a large budget, you can go out and buy a complete five-axis unit, such as the Pocket NC (currently, there is a V2 being offered), but for most hobbyists, the cost of something like this is going to be prohibitive.

This last video (the one that carves out the boat) is a great segue into talking about SCARA robots. As the robot is used for Assembly, the 'A' in SCARA is, of course, 'Assembly' as well as 'Articulated'. I once saw one of these, essentially a large robotic arm, fitted with a large 3D printer hot end and nozzle, 3D-print furniture right in front of me. Amazing! That's when I thought that I MUST make an arm like this to machine things. Needless to say, I never got around to it, but the previous video is half that equation. What if you have a fully articulated arm that looms over your workpiece and can go around it, cutting and drilling as it goes? It's not far-fetched to say that the CNC-er cutting the boat out of a large block of foam is a step in that direction.

SCARA robots are very interesting because they offer movement across the X, Y, and Z axes via an articulated arm, and mechanically, it is simple enough to add a way to rotate the toolhead in two more planes. The downside of robotic arms is that they do not offer the same structural rigidity as other machines. For example, the articulated arm can act like a big lever when it pushes against hard material if the toolhead is a spindle. However, for machining soft materials, such as foam, they present an interesting method to cut and carve, and I have even seen some used as 3D printers. To make the work area larger, you simply increase the size of the arm joints and the height of the Z axis. Rigidity can also be added by making the structure out of durable materials, such as metals (instead of plastics). Everything else is in the software.

Summary

We've covered a lot of material in this book and built several machines. We've also made use of other machines in our toolchain, including 3D printers, other CNC machines, and external machine shops, to effectively set up and utilize our tools. We learned about machines of varying complexity and explored some of the options available to us, not just for the hobbyist but for those who want to take CNC beyond the desktop.

While we have looked at several ways that you can CNC, we also have touched on some very complex means by which you can machine materials, using different concepts. Underneath all these mechanical approaches, the software and the concepts they use are consistent. All of them use some flavor of G-code that is generated by software that derives the toolpath from the base drawing.

As you explore your CNC aspirations, always consider looking ahead and exploring newer technologies. At a minimum, you should have some way to cut with a spindle and a laser. With just a 3018, even if equipped with a laser, you should be able to build bigger and better machines so that every generation that you build can fabricate the parts for the generation that follows. For maximum flexibility, add a suitable 3D printer, and you will have a well-equipped shop to make just about anything you like out of whatever materials you want to support.

I hope you have enjoyed our journey, and rather than say goodbye, I will just say, see you around the shop…

Index

‹packt›

www.packtpub.com

Subscribe to our online digital library for full access to over 7,000 books and videos, as well as industry leading tools to help you plan your personal development and advance your career. For more information, please visit our website.

Why subscribe?

- Spend less time learning and more time coding with practical eBooks and Videos from over 4,000 industry professionals

- Improve your learning with Skill Plans built especially for you

- Get a free eBook or video every month

- Fully searchable for easy access to vital information

- Copy and paste, print, and bookmark content

Did you know that Packt offers eBook versions of every book published, with PDF and ePub files available? You can upgrade to the eBook version at packtpub.com and as a print book customer, you are entitled to a discount on the eBook copy. Get in touch with us at customercare@packtpub.com for more details.

At www.packtpub.com, you can also read a collection of free technical articles, sign up for a range of free newsletters, and receive exclusive discounts and offers on Packt books and eBooks.

Other Books You May Enjoy

If you enjoyed this book, you may be interested in these other books by Packt:

Making Your CAM Journey Easier with Fusion 360

Fabrizio Cimò

ISBN: 978-1-80461-257-6

- Choose the best approach for different parts and shapes.
- Avoid design flaws from a manufacturing perspective.
- Discover the different machining strategies.
- Understand how different tool geometries can influence machining results.
- Discover how to check the tool simulation for errors.
- Understand possible fixtures for raw material blocks.
- Become proficient in optimizing parameters for your machine.
- Explore machining theory and formulas to evaluate cutting parameters.

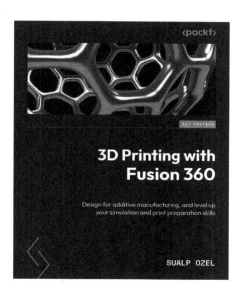

3D Printing with Fusion 360

Sualp Ozel

ISBN: 978-1-80324-664-2

- Use Autodesk Fusion to open, inspect, repair, and edit externally created designs for 3D printing
- Set up your 3D prints for different printing technologies, such as FFF, SLA/DLP, SLS, and MPBF
- Use templates to automate your additive operations, including part orientation, arrangement, and support
- Run process simulation for metal powder bed fusion and learn how to compensate for common print failure modes
- Optimize Fusion 360's preferences for 3D printing
- Export machine-specific file formats for 3D printing, such as G-Code, SLI, SLC, and CLI.

Packt is searching for authors like you

If you're interested in becoming an author for Packt, please visit `authors.packtpub.com` and apply today. We have worked with thousands of developers and tech professionals, just like you, to help them share their insight with the global tech community. You can make a general application, apply for a specific hot topic that we are recruiting an author for, or submit your own idea.

Hi!

I am Samer Najia, author of *A Tinker's Guide to CNC Basics*. I really hope you enjoyed reading this book and found it useful for increasing your productivity and efficiency using CNC.

It would really help me (and other potential readers!) if you could leave a review on Amazon sharing your thoughts on this book.

Go to the link below or scan the QR code to leave your review:

`https://packt.link/r/1803247495`

Your review will help us to understand what's worked well in this book, and what could be improved upon for future editions, so it really is appreciated.

Best Wishes,

Samer Najia

Download a free PDF copy of this book

Thanks for purchasing this book!

Do you like to read on the go but are unable to carry your print books everywhere?

Is your eBook purchase not compatible with the device of your choice?

Don't worry, now with every Packt book you get a DRM-free PDF version of that book at no cost.

Read anywhere, any place, on any device. Search, copy, and paste code from your favorite technical books directly into your application.

The perks don't stop there, you can get exclusive access to discounts, newsletters, and great free content in your inbox daily

Follow these simple steps to get the benefits:

1. Scan the QR code or visit the link below

https://packt.link/free-ebook/9781803247496

2. Submit your proof of purchase
3. That's it! We'll send your free PDF and other benefits to your email directly